Monographs

Series Editor: U. Veronesi

The European School of Oncology gratefully acknowledges Schering Plough International and its Essex and AESCA subsidiaries for an educational grant for the sponsorship of this task force and monograph.

K. Pummer (Ed.)

Biological Modulation of Solid Tumours by Interferons

With 12 Figures and 11 Tables

Springer-Verlag
Berlin Heidelberg New York
London Paris Tokyo
Hong Kong Barcelona
Budapest

Karl Pummer, M.D.

Universitätsklinik für Urologie
Auenbruggerplatz 1
8036 Graz, Austria

ISBN-13: 978-3-642-78868-0   e-ISBN-13: 978-3-642-78866-6
DOI: 10.1007/ 978-3-642-78866-6

Library of Congress Cataloging-in-Publication Data
Biological modulation of solid tumours by interferons / K. Pummer (ed.)
   (Monographs / European School of Oncology)
Includes bibliographical references.
   ISBN 3-540-57764-5 (alk. paper)
   ISBN 0-387-57764-5 (alk. paper)
1. Interferon--Therapeutic use. 2. Cancer--Adjuvant treatment. 3. Biological response modifiers. I. Pummer, K. (Karl), 1956-.
II. Series: Monographs (European School of Oncology) [DNLM: 1. Neoplasms--therapy. 2. Interferons--therapeutic use. 3.
Biological Response Modifiers. QZ 266 36153 1994] RC271.I46B54 1994 616.99'406--dc20 DNLM/DLC for Library of
Congress

Typesetting: Camera ready by editor
19/3130 - 5 4 3 2 1 0 — Printed on acid-free paper

Foreword

The European School of Oncology came into existence to respond to a need for information, education and training in the field of the diagnosis and treatment of cancer. There are two main reasons why such an initiative was considered necessary. Firstly, the teaching of oncology requires a rigorously multidisciplinary approach which is difficult for the Universities to put into practice since their system is mainly disciplinary orientated. Secondly, the rate of technological development that impinges on the diagnosis and treatment of cancer has been so rapid that it is not an easy task for medical faculties to adapt their curricula flexibly.

With its residential courses for organ pathologies and the seminars on new techniques (laser, monoclonal antibodies, imaging techniques etc.) or on the principal therapeutic controversies (conservative or mutilating surgery, primary or adjuvant chemotherapy, radiotherapy alone or integrated), it is the ambition of the European School of Oncology to fill a cultural and scientific gap and, thereby, create a bridge between the University and Industry and between these two and daily medical practice.

One of the more recent initiatives of ESO has been the institution of permanent study groups, also called task forces, where a limited number of leading experts are invited to meet once a year with the aim of defining the state of the art and possibly reaching a consensus on future developments in specific fields of oncology.

The ESO Monograph series was designed with the specific purpose of disseminating the results of these study group meetings, and providing concise and updated reviews of the topic discussed.

It was decided to keep the layout relatively simple, in order to restrict the costs and make the monographs available in the shortest possible time, thus overcoming a common problem in medical literature: that of the material being outdated even before publication.

Umberto Veronesi
Chairman Scientific Committee
European School of Oncology

Contents

Introduction

K. Pummer

Universitätsklinik für Urologie, Karl Franzens Universität, Auenbruggerplatz 1, 8036 Graz, Austria

Interferons were first discovered in 1957 by Isaacs and Lindenmann [1]. They are proteins produced by cells in response to virus infection, and they can confer resistance to attack by a wide range of viruses in a species-specific way. As soon as their importance in the regulation of immune responses as well as their involvement in the control of cell growth and differentiation had been recognised, the possibility of growth inhibition in cancer was suggested. Following the successful cloning of the interferon genes, in 1982 the first recombinant human interferons were produced on a commercial scale, allowing extensive research during the last ten years.

At present, highly suggestive evidence exists that interferons do have anticancer activity, although the range of responding tumours is limited. Tumours of haematopoietic or lymphatic origin are more likely to respond than solid tumours, and those responding may not be the most common ones. The activity in renal cell carcinoma and melanoma is controversial, the effects in ovarian or urothelial cancers appear to be promising, but the responses reported in the most common cancers such as breast cancer, colon cancer, non-small cell lung cell cancer and prostate cancer are disappointing.

The currently used classification of interferons is mainly based on their cellular origin and antigenicity. Human interferon-α, or leukocyte interferon, is produced by lymphoblasts, B lymphocytes and macrophages in response to viruses, foreign cells and B mitogens. Human interferon-β, or fibroblast interferon, is released from fibroblasts or epithelial cells stimulated with double-stranded ribonucleic acid. Human interferon-γ, or immune interferon, is produced by T lymphocytes upon induction by T mitogens or foreign antigens [2].

In order to exert regulatory activities on a cell, interaction of interferons with a specific cell surface receptor is necessary. The majority of the cells express two cell surface receptors, one shared by α- and β-interferon, the other for γ-interferon [3].

The maximum tolerated dose of interferons, which was assessed in a cohort of clinical phase I trials, mainly depends on the route of administration, the duration of therapy, and the performance status of the patients. For interferon-α, doses of up to 100×10^6 IU/m^2 have been shown to be safely administrable. Natural interferon-ß was administered up to 10×10^6 IU, whereas recombinant interferon-β was tolerated up to doses as high as 500×10^6 IU/m^2. For interferon-γ the maximum tolerated dose was assessed to be in the range of 2×10^5 to 2×10^8 IU/m^2. The relationship between dosage and response is still discussed and there is some evidence that the maximum tolerated dose is not necessarily the most effective one. Responses occur slowly, suggesting that in haematological diseases and solid tumours alike long-term administration of even very low doses might be superior to short-term high-dose schedules.

Major side effects seen in patients treated with various doses of interferons are headache, fever, chills, myalgia, anorexia, nausea and fatigue. At high doses, interferon-α exhibits severe central nervous system toxicity, although interferons do not cross the blood-brain barrier, thus suggesting that intermediates capable of crossing this blood-brain barrier are generated. Toxicity of interferon-γ is partially due to its activity on the endocrine system. It has been suggested that fatigue seen in patients receiving interferons might be caused by elevated corticosteroid levels [4]. Changes in lipid metabolism, particularly concerning triglyceride serum levels, have also been described [5]. In general, however, side effects are transient and hardly ever life-threatening.

The antitumour actions of interferons are multitudinous and since they form part of the cytokine network, things become even more complex, because *in vivo* generation of other cytokines by interferons is likely to occur. There are at least three ways in which interferons could affect tumour growth. Firstly, they can exert direct regulatory effects including growth inhibition or transcriptional inhibition on tumour cells. Secondly, with regard to their immunomodulatory capacity, they can enhance or even initiate host response. Thirdly, regulatory effects, other than immunological and probably yet unknown, on host/tumour relationship might play an important role. Because of their pleiotropic actions on cells, different mechanisms may apply to different tumours.

In order to provide a state-of-the-art on the biological modulations of interferons on selected solid tumours, an international panel of experts has been formed. In preparing this monograph, the editor was fortunate to collaborate with this outstanding group of clinicians and scientists, and wishes to express his warmest appreciation.

REFERENCES

1   Isaacs A, Lindenmann J: Virus interference. I. The interferon. Proceedings of the Royal Society of London (Series B) 1957: 259-267

2   Horoszewicz JS, Murphy GP: An assessment of the current use of human interferons in therapy of urological cancers. J Urol 1989 (142): 1173-1180

3   Rubinstein M , Orchansky P: The interferon receptors. CRC Critical Reviews in Biochemistry 1986 (21): 249-275

4   Balkwill FR: Side-effects of interferon therapy. In: Balkwill FR (ed) Cytokines in Cancer Therapy. Oxford University Press, New York 1989 pp 44-45

5   Kurzrock R, Quesada JR, Rosenblum MG, Sherwin SA, Gutterman JU: Phase I study of i.v. administered γ-interferon in cancer patients. Cancer Treatment Reports 1986 (70): 1357-1363

The Interferons: Basic Concepts Concerning their Modulatory Effects on the Immunological System

M. Alvarez-Mon [1], J. Keller [1], L. Molto [1], L. Manzano [1], E. Reyes [1], M. Rodriguez-Zapata[1] J. Carballido [2], and S. Vaquer [3]

1   Department of Medicine, University Hospital "Principe de Asturias", Facultad de Medicina, 28871 Alcalá de Henares
2   Hospital Universitario de Guadalajara, Department of Urology, Clinica Puerta de Hierro, 28034 Madrid
3   Department of Gynaecology, Hospital General de Guadalajara, Universidad de Alcalá, 28871 Alcalá de Henares, Spain

The immunological system is composed of diverse molecules and an intricate matrix of cells equipped with a variety of biological functions that react to a multitude of antigenic assaults to which the human organism is exposed. The most essential and characteristic distinction of this system is the unique capacity that some of its components possess which translates into recognition in a very specific way of given molecular fragments or antigens. This special ability provides for the proper physical and chemical interaction among different antigens with the specific antigenic or clonotypic receptors located on the surface of the T lymphocytes, with immunoglobulins located on the cytoplasmic membrane of the B cells, or with those dispersed in the extracellular space. This specific recognition induced by the interaction between the antigen and the clonotypic receptor, or immunoglobulin, leads to the activation of the corresponding T or B lymphocytic clone, respectively. The proliferation, expansion and maturing of these specifically activated lymphocytic clones are responsible for encouraging immune response against the inducer (antigen). In turn, these cellular processes are produced and regulated by the series of antigen nonspecific cytokines which include fundamental molecular compounds such as lymphokines and monokines.

The cytokines in question are therefore involved in the creation of a precise response in face of antigen threat and play a vital role in programming the defences of the organism which have been affected by the neoplasia. It should also be mentioned that these cytokines may have a direct influence on the growth and differentiation of tumoural cells, or on those which are invaded by microorganisms, or even on those very same invaders. At present, a wide range of molecules have been classified and grouped under the name of cytokines. One of the best known groups which in biochemical terms has undergone intense investigation and is consequently widely researched is the group of interferons. The biological activities of interferons and their clinical applications are the subject of continuous studies and research.

The protein capable of inducing cell resistance in response to viral aggression, and therefore called interferon (IFN), was discovered by Isaac and Lindenman in 1957. Much exhaustive effort has been put into the search for more information concerning this molecular family, in particular with regard to aspects of its molecular character and its actions on different cellular systems.

Types of Interferon

IFNs were initially described as virus-induced proteins. Nevertheless, experimental evidence leads us to believe that IFNs may also be constituents, as well as possibly continuously induced and synthesised products. A quantified *in vivo* decline in the basal levels of the enzyme 2'-5'-oligoadenylate synthetase, which is almost exclusively induced by IFNs, immediately after *in vivo* administration of antibodies against these IFNs, has been a major proof of this assumption.

IFN synthesis can be verified in the infected experimental animals because, if there is enough of this synthesised material in the samples of serum from experimental animals, it will correspond to a clinical condition known as interferonaemia. Several microorganisms have also been involved in the *in vivo* synthesis of IFNs. These include viruses and bacteria as well as chlamydias richettsiae, mycoplasma, protozoas, and fungii, as well as their natural products. It has also been found (in experimental animal models) that *in vivo* administration of synthetic compounds such as polynucleotides, can induce production of IFNs. Several mitogens were found to be IFN inducers as well.

IFNs were initially classified on the basis of different criteria such as cellular types that synthesise them, their chemical characteristics, and their antigenic properties. At present, and thanks largely to progress made in the cloning of different genes that codify various polypeptides constituting this molecular family, it is possible to identify 3 classes of IFNs, denominated alpha, beta and gamma [1,2]. The type and *in vivo* production of any of these depends on the inducing stimulus as well as on the characteristics of the cell which reacts to this stimulus [3].

IFN-α, in which α2 is the predominant type, is produced mainly by T lymphocytes, B lymphocytes, NK cells, monocytes as well as by macrophages and some granulocytes and leukocytes. On the other hand, IFN-β is produced by fibroblasts and epithelial cells as well as by some immune cells. These two types of molecules form the class I interferons.

IFN-γ, also called immune IFN, belongs to class II and is synthesised by T lymphocytes and NK cells in the course of immunological responses. Thus, it is produced through the accessory stimulus provided by the antigen presenting cells to T lymphocytes, or through the actions of soluble regulating factors of the immunological system (cytokines) which exert an influence on these T lymphocytes.

IFN-α and β share at least 30% of homology in their primary structure, whereas the amino-acid sequence of IFN-γ has no relation whatsoever with the two afore-mentioned types of interfer-

ons. IFN-α constitutes a molecular family of at least 24 different species, codified by the corresponding genes localised at the ninth chromosome. The molecular weight of these IFNs varies between 16,000 and 27,000 daltons and they have a defined homology of aminoacids of up to 50% among different species. They show persistent similarities in the marked regions 139 and 157. IFN-α in positions 1, 29, 98/99 and 138/139, 4 cisteins with 2 disulphur bonds between the first and the third and between the second and the fourth are to be found. The secondary structure of these proteins is determined by the folds of an α-helix. In general, IFN-α is not found to be glycosylated but stable at pH 2, nevertheless the existence of species that undergo lysis in the acidic medium has been confirmed.

IFN-β includes 2 proteinic species which are completely heterogeneous and are codified by two different genes located at the ninth chromosome. As a matter of fact, IFN-β was the first to be discovered and thoroughly studied. Its molecular weight is 20,000 daltons and it possesses 3 cisteins in its primary structure located in the 17, 31, 141 positions with the disulphur bond between the last two. The predominating secondary structure of this species is that of the α-helix. IFN-β is found to be glycosylated and, as IFN-α, is stable in the acidic medium of pH 2.

IFN-γ, on the other hand, is a glycoprotein consisting of 143 amino-acids whose synthesis is coded by a gene located at the twelfth chromosome. Till now, two different molecular forms have been classified, one with a molecular weight of 20,000 daltons and the other of 25,000 daltons, differentiated from each other by the level of glycosylation. These proteins lack the cistein residues and the disulphur bonds are missing. Unlike the other classes of IFNs, IFN-γ undergoes lysis in the acidic medium. These findings have been employed in the labelling of different brands of IFNs [4].

Actions

IFNs are substances with multiple pleiotropic actions, produced in response to any kind of inflammatory or infectious stimulus and are capable, as such, of regulating the synthesis of nucleic acids and proteins and modifying the processes of growth, differentiation and cellular activation [5-8].

In the same way as other cytokines, IFNs express their biological activity by means of interaction with precisely defined receptors located in the cytoplasmic membrane (9). Thus the cells that lack these receptors are not affected by the biological action of INFs. As a result, the existence of these specific receptors conditions a proper immunological function. Experimental evidence exists today which allows us to assume that IFN-α and β share the same receptor, while γ attaches itself to a completely different one. The gene that codifies for IFN-α and β receptors is found on chromosome 21.

The binding of IFN-α and β to their receptors is saturable with a dissociation being constant of between $1 \cdot 10^{(-9)}$ and $1 \cdot 10^{(-11)}$ M. The number of receptors per cell falls between 200 and 6,000. Other receptors of low affinity have been described, whose number in the cellular membrane could be much higher. The number of receptors per cell for IFN-γ varies between 1,000 and 10,000 with a dissociation of $1 \cdot 10^{(-9)}$ and $1 \cdot 10^{(-11)}$ M. It seems that interaction of these IFNs with their receptors is not sufficient enough to induce a desired biological response but it is conditioned by the internalisation. Although some contradictory publications do exist, it was not possible to demonstrate competition for their original receptors among IFN-α and β and IFN-γ.

Little is known about the molecular processes involved in the interaction of IFNs with their specific receptors. Evidently, their biological actions are triggered by the induction of the enzymatic systems that are capable of inhibiting the proteinic translocation, as well as the action of the protein kinases. Other metabolic pathways that modify the synthesis of DNA, RNA and proteins are also modulated by the above-mentioned systems. These combined biochemical effects are the principal mechanisms governing the main biological actions attributed to IFNs. Such action consists in the induction of resistance to viral replication and the regulation of the cellular proliferation and differentiation processes. IFNs are also elaborated in the course of an immunological response and regulate and stimulate these enzymatic reactions which lead to activation, proliferation and differentiation of the immune cells. There is a cer-

tain diversity of biological actions among different classes of IFNs, as well as among different subtypes of the same class. In addition, the same biological effect of induction has been observed in identical subtypes of IFNs. The activity of cytotoxic effector lymphocytes including T cells and NK cells, as well as that of macrophages can be enhanced by IFNs. These different effector cells activated in the course of immunological response and the antibodies produced by B lymphocytes can be responsible for cytotoxic activity against virus infected cells and tumoural cells. Thus, these activated effector cells can suppress the progression of the infection and destroy the tumour cells or retard the growth and systemic dissemination of neoplasia [10-16].

The biological activity of IFNs is basically due to their ability to regulate specific genes. In most cases, the effects of IFNs may influence gene expression activation, although regression takes place in some proteins. IFNs activate a number of genes, some of which have approximately a 30-base pair (bp) consensus sequence present in their 5' flanking region. This sequence acts like an IFN-senstive amplifier. Among the proteins induced by the treatment with IFNs, the most extensively studied is the synthetase group. They are composed of several synthetases that form part of a complex system which also includes an endokinase requiring the product of the synthetase 2'-5' oligoadenylate (2'5'A) for its activity. The activated nuclease is responsible for some of the biological effects of IFNs. This enzyme (2'5'-A-activated endonuclease) is a constituent organic compound in some cells, but in others it is induced by IFN treatment. There are several systems in which antiviral activity appears to be related to an activated endonuclease. The activated endonuclease most likely produces its effects by cleaving ribosomal RNA in intact ribosomes and hydrolises messenger RNA with a subsequent halt of protein synthesis. A ribosome-associated protein kinase is autophosphorylated by direct contact with dsRNA and ATP. IFNs can also inhibit the replication of retroviruses and some other membrane-associated viruses. In some instances, IFNs do not influence the synthesis of viral proteins or RNA at all, but they modify the assembly of mature virons. Viral particles due to the faulty assembling process are capable of composing the plasma membrane but are unable to free themselves from the cell surface. They may also lack the glycoprotein which is necessary for viral absorption.

IFNs also have the ability to inhibit growth of both normal and transformed cells. One of the additional functions of IFNs is to counter the actions of several platelet growth inducing factors (platelet derived growth factor) and to impede the action of epidermal growth factor as well. IFN-α, in addition, can inhibit the expression of various oncogenes. IFN-α has a negative regulatory effect on c-myc expression at a post-transcriptional level. Apart from the already-mentioned actions, IFNs can also mediate a down-regulation of c-ha-ras at the level of transcription. It has also been shown that IFN-α can retard cellular transformation. In fact, there have been a number of reports that suggest the possibility of IFNs being able to reverse the transformed and tumourigenic phenotype in several *in vitro* cell lines.

In general, the biological action of IFNs is initiated by the binding of these molecules to their specific receptors in the cytoplasmic membrane which is then followed by internalisation, as previously mentioned. This determines the initiation of a series of biochemical processes with activation and/or inhibition of diverse enzymatic systems which cause modifications in the synthesis of DNA, RNA and proteins, as well as significant alternations in metabolic pathways. These biochemical and genetic effects constitute the basis of biological actions associated with IFNs. The obvious conclusion resulting from this brief revision leads us to assume that the biological importance of IFNs is relevant overall and that they also possess a crucial role in regulating the immune system. This phenomenon in turn has an infinite number of practical and therapeutic applications [17,18].

The Immunomodulator Effects of IFNs

An understanding of the potential therapeutic use of IFNs can be acquired from the analysis of their immunomodulatory effects. The cellular compartment of the immune system is composed of T lymphocytes, B lymphocytes, NK cells and the so-called accessory cells that include macrophages, monocytes, dendritic cells and Langherhan cells, respectively. The molecular component is made up of antibodies, a complement system and a wide range of molecules or cytokines which are called monokines or lymphokines, depending on the type of cells that secrete them, i.e., macrophages or lymphocytes. It is quite evident that IFNs are a part of this intrinsic molecular complex [19].

As previously described, the specific recognizing capacity of the immune system is determined by two basic types of molecules; the immunoglobulins and the clonotypic receptors of T lymphocytes. These antigen receptors are equipped with great structural variability, expressing immeasurable amounts of slightly different forms. This molecular diversity constitutes the pool of specific forms within an individual, capable of recognising with greater or lesser affinity all possible antigenic conformations existing in nature. These structures share a similar three-dimensional configuration, although the strategies of recognition and activation of both lymphocytic populations of T and B are completely different. Thus, while the B lymphocytes can activate themselves in the presence of soluble antigens, the T cells require for their activation recognition of the antigen on the cytoplasmic membranes in the presence of molecules of the main histocompatibility system [20-24].

The genetic products of the human histocompatibility system which determine the immunological identity of the individual, include the denominated class I molecules present in the cytoplasmic membrane of the nucleated cells of the organism. The class II molecules of the main human histocompatibility system are expressed on the cytoplasmic membrane of the accessory cells, B lymphocytes and activated T lymphocytes which are functional only under proper physiological conditions. However, in several diseases these class II molecules can appear on the cytoplasmic membrane of other cells. Therefore, recognition of the antigen by T lymphocytes is limited by the molecules of the major histocompatibility complex. The CD4+ T lymphocyte antigen recognition is restricted by class II molecules of the major histocompatibility complex. It follows that the CD8+ T lymphocyte antigen recognition is restricted by class I molecules of the same major histocompatibility complex.

The lymphocytes of the immune system are

activated through interaction with an antigen following a determined order and are selective to those clones with receptors in their cytoplasmic membrane that can specifically recognise the given antigen with greater affinity. However, subsequent proliferation and cellular differentiation are regulated by antigen non-specific molecules (cytokines) secreted by T and B lymphocytes, NK cells and macrophages.

Thus, the disabling of the antigen by the macrophage and its subsequent presentation to and interaction with T-helper lymphocytes initiates a process of release of different molecules, such as interleukin 1, tumour necrosis factor (TNF), IFN-α, etc into the medium. These monokines act upon those cells that express their specific receptors in their cytoplasmic membranes. After interleukin 1 has bound with its receptor located on the surface of an activated T-helper lymphocyte, it begins to secrete diverse lymphokines such as interleukin 2, interleukin 4, interleukin 6, IFN-α and γ, transforming growth factor β, lymphotoxin, etc. These different molecules, in turn, act on cells that express corresponding receptors. In this way, interleukin 2 interacts with the cytotoxic T lymphocytes which are activated after recognising a given antigen on the cytoplasmic membrane in the presence of the class I molecules of the major histocompatibility complex. After interaction of the interleukin 2 receptor, proliferation of the T helper and cytotoxic T lymphocytes takes place. The capacity for lysis of these cytotoxic T lymphocytes is also enhanced. In addition, the lymphokines secreted by T-helper lymphocytes are capable of regulating the proliferation and differentiation of B lymphocytes towards plasma cells and the secretion of immunoglobulins. Thus, clonal B lymphocyte activation is initiated by the interaction of the surface immunoglobulins with soluble antigen or by direct contact with antigens present on the cytoplasmic membrane of macrophages. However, contrary to the T lymphocytes, the proliferation and differentation of these clonal activated B lymphocytes are regulated by non-antigen-specific monokines and lymphokines [25-29].

When regulating these distinct processes of activation, proliferation and differentiation of T lymphocytes, B lymphocytes, NK cells and macrophages, IFNs exert an important function, either by a direct action on the target immune cell, or indirectly by regulating the secretion of different monokines and cytokines with a direct effect on those cells.

Among the direct immunostimulating effects of IFNs observed in experimental models and in *in vitro* assays, there is an increase of the effector capacity of the monocytes and macrophages against tumoural or infected cells seen in these molecules. On the other hand, IFNs have the capacity of inducing lytic activity in NK cells and in CD8+ lymphocytes. They are also capable of spawning a family of LAK cells, whether through direct action or in association with interleukin 2. They also increase the cytotoxic capacity of T lymphocytes and NK cells that were previously stimulated by interleukin 2. They continuously synergise with this lymphokine, thereby enhancing its cytotoxic activity to the level qualitatively similar to that obtained with IL-2 alone. Acting on NK cells, they can increase their capacity of lysis through antibodies (antibody-dependent cytotoxicity). Unlike T lymphocytes, NK cells possess a kind of cytotoxic activity that can induce lysis of tumoural cells or cells infected by virus without the need of previous immunological recognition. Antigenic recognition restriction by molecules of the main histocompatibility system is also absent in these NK cells. LAK cells represent a new model of non-specific cytotoxic activity exerted by the CMNSP which has lytic capacity against fresh tumoural cells. These cells are also autologous and show resistance to lysis by the conventional NK cells. This evidence, which surely heralds new clinical applications, has been observed in patients with advanced forms of neoplasia where some forms of rejection against tumoural cells were patent. In some experimental models the modulatory capacity of IFN is basically related to the increased lytic activity of the existing cytotoxic cells. IFN-activated NK cells express a relative speed-up in the formation of the lytic conjugates with target cells. This cytotoxic enhancer effect is dependent on the dosage of IFN and on its incubation time. Likewise, the effect of IFNs upon the NK activity can also be found in the presence of synthetic inductors of IFN. Paradoxically, in some occasions IFNs can induce resistence in tumoural cells sensitive to lysis by way of the effector NK cells. But the majority of the tumour, or virus-infected cells are insensitive to this protective effect. However, IFNs may sometimes increase the margin of reactivity of the tumour cells against the host's immunological system. Among these protective effects induced by the secretion and/or interaction with different cytotoxins, it was noted that IFNs are capable of increasing the amounts of interleukin 1 and 2. This phenomenon also applies to the tumour necrosis factor and to some soluble forms of receptors for IL-2 [30-36].

IFN-α and β can also induce differentiation of pre-NK cells to acquire fully cytolytic forms. Administration of IFNs *in vivo* and *in vitro* produces a significant increase in NK cell activity. It has been demonstrated that IFN-α can enhance depressed NK activity of peripheral blood mononuclear cells from patients with advanced solid tumours. We have ascertained that IFN-α can increase the diminished cytotoxic activity of peripheral blood mononuclear cells from patients with different solid tumours such as infiltrating transitional cell carcinoma of the bladder, head and neck cancers and cancer of the uterine cervix (Fig. 1). The cytotoxic inducer effect of IFN-α upon the depressed NK activity of the peripheral blood mononuclear cells from these patients is both time and dose dependent (Fig. 2 and 3). Furthermore, the en-

hancing effect of IFN-α is not associated with the expansion of the effector NK cells. As already mentioned, this specific action of IFN-α is attributed to the increase in NK cell activity. The regulatory effect of IFN-α on NK cells can be supplementary to the action of other cytokines such as interleukin 2. We have ascertained that cytotoxic inducer effect of interleukin 2 can be enhanced by IFN-α in peripheral blood cells from patients with renal cancer and with glioblastoma multiforme. These enhancing effects of IFN-α on the modulatory action of interleukin 2 can be demonstrated *in vitro* by the simultaneous or sequential addition of both cytokines to the cellular medium culture. We have also noticed this synergistic effect after *in vivo* administration of interleukin 2 and IFN-α2b to patients with disseminated renal cancer. The cytotoxic activity exhibited by IFN-α in activated peripheral mononuclear cells from patients with solid tumours can be directed against NK-sensitive and NK-resistant target cells. This effect of IFN-α on the cytotoxic activity of NK cells is observed in different anatomical compartments. We have shown

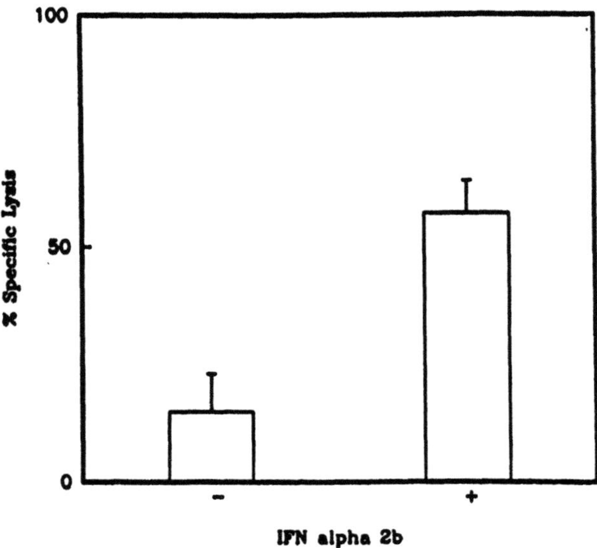

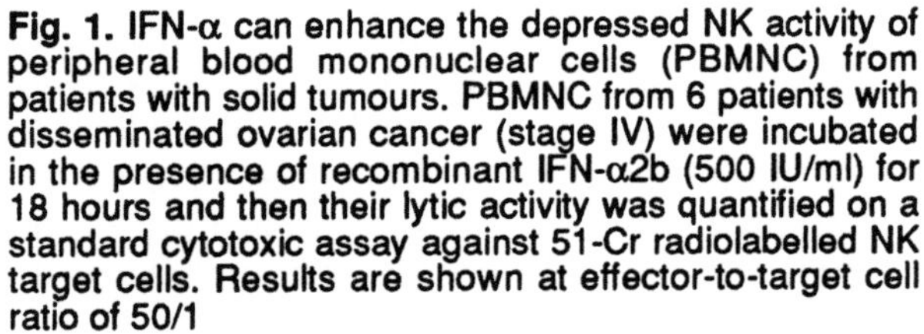

Fig. 1. IFN-α can enhance the depressed NK activity of peripheral blood mononuclear cells (PBMNC) from patients with solid tumours. PBMNC from 6 patients with disseminated ovarian cancer (stage IV) were incubated in the presence of recombinant IFN-α2b (500 IU/ml) for 18 hours and then their lytic activity was quantified on a standard cytotoxic assay against 51-Cr radiolabelled NK target cells. Results are shown at effector-to-target cell ratio of 50/1

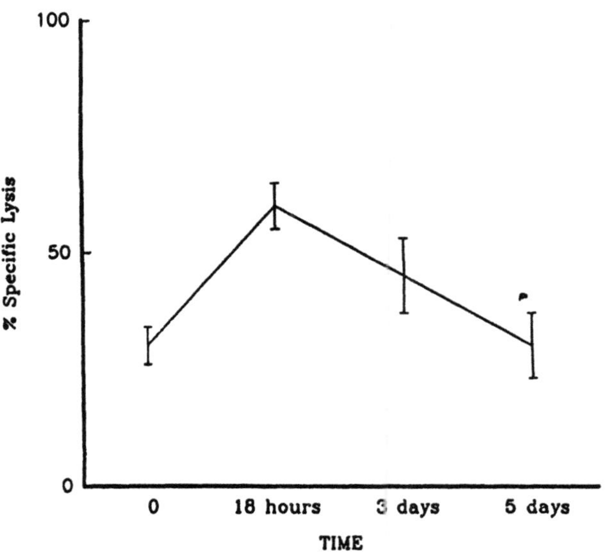

Fig. 2. IFN-α enhances the NK activity of peripheral blood mononuclear cells (PBMNC) in a time-dependent manner. PBMNC from 7 patients with disseminated renal cancer were incubated in the presence of recombinant IFN-α2b (500 IU/ml) for 18 hours, 3 and 5 days, and their lytic activity was quantified on a standard cytotoxic activity assay against 51-Cr radiolabelled NK target cells. Results are shown at an effector-to-target cell ratio of 50/1

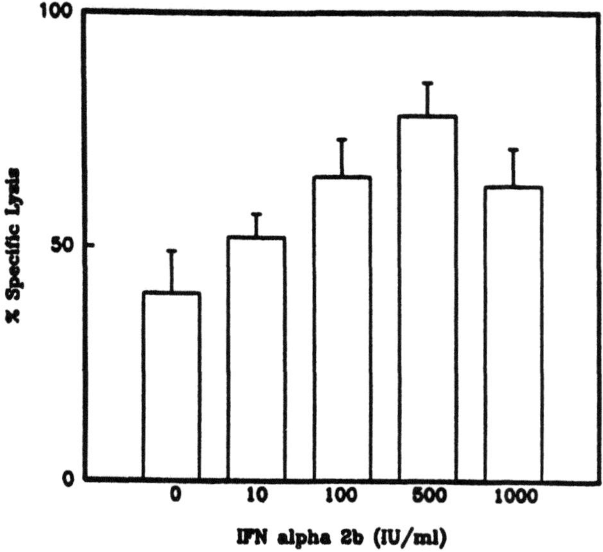

Fig. 3. IFN-α can enhance the activity of peripheral blood mononuclear cells (PBMNC) in a dose-dependent manner. PBMNC from patients with colorectal adenocarcinoma were incubated in the presence of different concentrations of recombinant IFN-α2b (1,000, 500, 100, and 10 IU/ml) for 18 hours and their lytic activity was quantified on standard cytotoxic assay against 51-Cr radiolabelled NK target cells. Results are shown at an effector-to-target cell ratio of 50/1

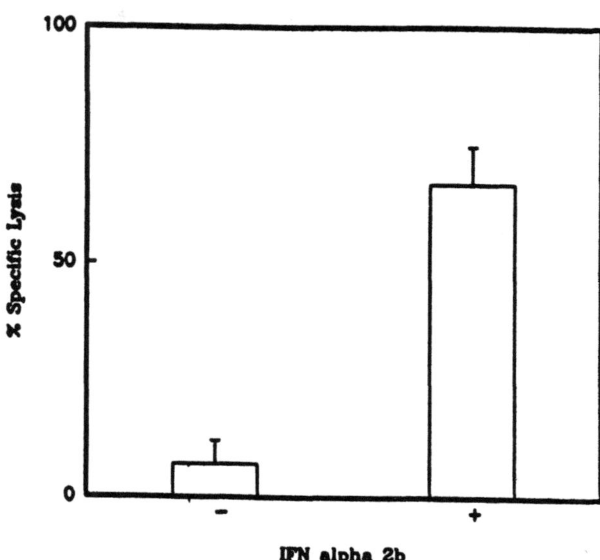

Fig. 4. IFN-α can enhance the NK activity of mononuclear cells from ascitic fluid. Mononuclear cells from 6 patients with peritoneal disseminated ovarian cancer were incubated in the presence of recombinant IFN-α2b (500 IU/ml) for 18 hours, and their lytic activity was quantified on standard cytotoxic assay against 51-Cr radiolabelled NK target cells. Results are shown at an effector-to-target cell ratio of 50/1

that IFN-α can enhance decreased NK activity of mononuclear cells from ascitic fluid in patients with disseminated ovarian carcinoma (Fig. 4). We have also noticed the cytotoxic inducer effect of IFN-α on mononuclear cells from regional lymph nodes of patients with solid tumours such as breast cancer. This NK activity inducer effect of IFN-α can be demonstrated *in vivo* after administration of this cytokine. It has been shown that prophylactic intravesical administration of IFN-α to patients with superficial cell carcinoma of the bladder drastically increases the NK activity on peripheral blood mononuclear cells [37,38].

Among the different immunomodulator actions induced by IFN, it has been demonstrated that IFN-α and β in macrophages are capable of increasing the release of interleukin 1 and TNF-α. Evidently, the rise of this monokine amplifies the immune response by interacting with T-helper lymphocytes. The higher production of TNF-α can also play a role in the lysis of neoplastic cells and in the activation, proliferation and differentiation of T and B lymphocytes. Similarly, both IFNs increase the cytotoxic effect of the above-mentioned macrophages on

the tumoural cells and participate in the lysis of the intercellular organisms and in the lysis of virus-infected cells. It should also be noted that IFNs are capable of increasing the release of interleukin 2 as well as the multiplication of soluble forms of receptors for IL-2 , which in turn affects the regulation of T lymphocytes. In the presence of different cellular signals, IFNs can have diverse effects on B lymphocytes. It has been reported that they can either enhance or suppress the pro-liferation and differentiation of B lymphocytes, and it has also been shown that IFNs can increase the expression of immunoglobulin-specific receptors.

IFNs have additional immunomodulatory effects such as the regulation of the expression of the major histocompatibility complex. IFN-γ increases not only the expression of the class I molecules in all cells in a similar way to that of IFN-α and β, but it also increases the expression of the class II molecules in the accessory cells and subsequently enhances its accessory function.

The different effects of IFNs depend on the type of IFN considered, their concentration and the time of exposure, as well as the degree of sensitivity of the assayed receptor cell. It is evident that the system of IFNs forms a complex and intrinsic molecular pool which interacts with the rest of the monokines and lymphokines. It can be altered quantitively/ qualitatively by different mechanisms. Thus some proteins of viral origin are capable of obstructing intracellular modifications brought about by IFNs, such as the expression of molecules of the major histocompatibility complex and of viral antigens on the cellular surface. Macrophages from peripheral blood can also, under strict assay conditions, inhibit or amplify through direct contact, reponse of NK cells produced by IFNs. Alterations in the synthesis or response of IFNs can be responsible for the pathogenesis of several infectious and tumour diseases. The local growth and the systemic dissemination of the tumours are associated with an inadequate immunological response to tumour cells. Thus, the enhancement and/or restoration of the activity of the different immune effector cells can serve as a therapeutic option in these diseases. In this sense, the therapeutic administration of IFNs has a substitutive significance in several viral and tumour diseases. An optimal application of these cytokines may be by way of regular minimum doses, rather than by high doses at intervals.

REFERENCES

1   Pestka S, Langer AJ, Zoon KC, Samuel CE: Interferons and their actions. Ann Rev Biochem 1987 (56):727-777

2   Zoon KC: Human interferons, structure and function. Int 1988 (9):1-11

3   Hooks JJ, Detrick B: Evaluation of the interferon system. In: Rose NR, de Macario EC, Fahey JL, Friedman H, Penn GM (eds) Manual of Clinical Laboratory Immunology. American Society for Microbiology, Washington 1992 pp 240-243

4   Balkwill FR: Interferons. Lancet 1989 (i):1060-1063

5   Le J, Prensky W, Yip KY, Chang Z, Hoffman T, Stevenson HC, Balasz I, Sadlik JR, Vilceck J: Activation of human monocyte cytotoxicity by natural and recombinant immune interferon. J Immunol 1983 (131):2821

6   Stark GR, Kerr IM: Interferon-dependent signaling pathways: DNA elements, transcription factors, mutations and effects of viral proteins. J Interferon Res 1992 (12):147-151

7   Overall ML, Chambers P, Hertzog PJ: Different interactions of interferon alpha subtypes at the surface of epithelial and lymphoid cells. J Interferon Res 1992 (12):281-288

8   Cebrian M, Yagüe E, de Landazuri MO, Rodriguez-Moya M, Fresno M, Pezzi N, Llamazares S, Sanchez-Madrid F: Different functional sites of rIFN alpha 2 and their relation to the cellular receptor binding site. J Immunol 1987 (138):484-490

9   Aguet M, Mogense KE: Interferons receptors. Interferon 1984 (5):1-19

10 De Maeyer-Guignard J, de Maeyer E: Immunomodulation by interferons: Recent developments. Interferons 1985 (6):69-91

11 Rossi GB: Interferons and cell differentiation. Interferon 1985

12 Bancherau J: Anitiviral properties of interferons. J Pharm Clin 1986 (5):193-200

13 Fleischmann CM, Fleischmann WR: Differential antiproliferative activities of IFNs alpha, beta and gamma: kinetics of establishment of their antiproliferative effects and the rapid development of resistance to IFNs alpha and beta. J Biol Reg & Hom Agents 1988 (2):173-185

14 Carballido J, Moltò L, Manzano L, Olivier C; Salmeròn O, Alvarez-Mon A: Interferon alpha 2b enhances the natural killer activity of patients with transitional cell carcinoma of the bladder. Cancer (in press)

15 Jonak GJ, Knight EJ: Interferons and the Regulation of Oncogens. Interferons 1986 (7):167-185

16 Kimchi A: Autocrine interferon and the suppression of the C-myc nuclear oncogene. Interferons 1987 (8):86-110

17 Alvarez-Mon M, Durantez A: Fisiopatologia de las linfocinas y las monocinas. Med Clin 1987 (89):387-393

18 Friedman RM, Vogel SN: Interferons with special emphasis on the immune system. Adv Immunol 1983 (34):97-140

19 Alvarez-Mon M: Interleucinas, en avances en medicina interna. Ed Arn 1988 ():233-239

20 Acuto O, Reinherz EL: The human T-cell receptor. N Engl J Med 1985 (312):1100-1111

21 Jeske DJ, Capra JD: Immunoglobulins: Structure and function. In: Paul WE (ed) Fundamental Immunology. Raven Press, New York 1984 pp 131-165

22 Doutinho A, Forni L, Homlberg D, Ivars F, Vaz N: From an antigen-centered, clonal perspective of immune responses to an organism-centered, network perspective of immune autonomous activity in a self-referential immune system. Immunol Rev 1984 (79):151-168

23 Alvarez-Mon M: Bases immunologicas del tratamiento del rechazo. En: Ministerio de Sanidad y Consumo (ed) De la Donacion del Transplante. 1987 pp 27-39

24 Alvarez-Mon M, Kehrl JH, Fauci AS: A potential role for adrenocorticotropin in regulating human B lymphocyte functions. J Immunol 1985 (135):3823-3826

25 Sancho L, de la Hera A, Nogalies A, Martinez C, Alvarez-Mon M: Induction of NK activity in newborns. N Engl J Med 1986 (314):57-58

26 Kehrl JH, Wakefield LM, Roberts AB, Jakowlen S, Alvarez-Mon M, Derynck R, Sporn MB, Fauci AS: The production of TGF-β by the human T lymphocytes and its potential role in regulation of T cell growth. J Exo Med 1986 (163):1037-1050

27 Vaquer S, de la Hera A, Jorda J, Martinez AC, Escudero M, Alvarez-Mon M: Diminished natural killer activity in pregnancy: modulation of NK activity by interleukin 2 and interferon. Scand J Immunol 1987 (26):691-698

28 Kehrl JH, Alvarez-Mon M, Delsing GA, Fauci AS: Lymphokine is an important T cell derived growth factor for human B cells. Science 1987 (238):1144-1146

29 Gaspar ML, Alvarez-Mon M, Gutierez CM: Cell activiation pathway in human systemic lupus erythematosos (SLE); imbalanced in vitro production of lymphokines and association with serum analytical findings. J Clin Immunol 1988 (6):266-274

30 Mogensen SC, Virelizier JC: The interferon macrophage allience. Interferon 1987 (8):55-84

31 Billiau A: The interferon-interleukin I connection. Interferon 1987 (9):91-111

32 Bonilla F, Alvarez-Mon M, Merino F, De la Hera A, Ales JE, Durantez A: Interleukin 2 induces cytotoxic activity in lymphocytes from regional axiliary nodes of breast cancer patients. Cancer 1988 (61):629-634

33 Trinchieri G, Santoli D, Granato D, Perussia B: Antagonistic effects of interferons on the cytotoxicity mediated by natural killer cells. Fed Proc 1981 (40):2705-2710

34 Holán V, Kohno K, Minowada J: Natural human interferon alpha augments interleukin-2 production by a direct action on the activated IL-2 producing T cells? J Interferon Res 1991 (11):319-325

35 Onji M, Yamaguchi S, Masumoto T, Fazle Akbar SM, Kajino K, Ohta Y: Induction of the interleukin-2 receptor (p55) by interferons? J Interferon Res 1991 (11):237-241

36 Foster GR, Ackrill AM, Goldin RD, Kerr IM, Thomas

HC, Stark GR: Expression of the terminal protein region of hepatitis B virus inhibits cellular responses to interferons alpha and gamma and double-stranded RNA. Proc Natl Acad Sci 1991 (88):2888-2892

37 Alvarez-Mon M, Moltó LM, Manzano L, Olivier C, Carballido JA: Immunomodulatory effect of interferon alpha 2b on natural killer cells and T lymphocytes from patients with transitional cell carcinoma of the bladder. Anticancer Drugs 1992 (3):5-8

38 Carballido JA, Moltó LM, Olivier C, Manzano L, Alvarez-Mon M: Analysis of the effect of intravesical treatment with interferon alpha 2b on   the clinical evolution and on the in vivo function of T lymphocytes and natural killer cells in patients with superficial bladder tumors. Anticancer Drugs 1992 (3):9-12

39 Lapeña P, Isasi C, Moltó L, Martinez R, Vaquero J, Alvarez-Mon M: Interleukin 2 and interferon alpha modulation of the lymphocyte non-major histocompatibility-restricted lytic activity in glioblastoma patients. INT J Immunopharmacol 1992 (14):1307-1313

Biomodulation of 5-Fluorouracil by Interferon-Alpha

H. Stöger and H. Samonigg

Department of Medicine, Division of Oncology, Karl Franzens University , Auenbruggerplatz 15, 8036 Graz, Austria

The fluorinated pyrimidine, 5-fluorouracil (5-FU), is the most commonly employed agent against tumours of the gastrointestinal tract. Nevertheless, it offers little benefit as a single agent. Response rates in colon cancer, for example, range from 7% to 25%. When patients do achieve an objective response, the duration is generally short, and complete responses are uncommon.

The wide variation in reported response rates for 5-FU therapy in metastatic colorectal cancer can be attributed to variables in the patient groups, differences in methods of response assessment, and variability in response criteria. Taking such factors into consideration, the activity of all these treatments is limited to a response rate of approximately 20%, with no significant impact on survival.

Combinations of 5-FU and other chemotherapeutic agents are rarely better than single-agent 5-FU [1]. One exception is FAM (5-FU, doxorubicin and mitomycin) in gastric cancer. Even with this combination of 3 active drugs, the objective response rates are approximately 40% and median survival for patients with metastatic disease treated with this regimen is about 29 weeks [2].

One theoretical approach to improving the response rates of 5-FU is to intensify the dose. This strategy is limited in practice, however, by a severe mucositis that develops with high-dose 5-FU therapy [3]. Therefore, other strategies have been attempted. These include locoregional therapy with either hepatic artery infusion or intraperitoneal treatments, prolonged systemic infusions with 5-FU, or modulation of the effects of 5-FU with non-toxic agents.

The modulation of 5-FU activity with non-toxic agents, called biochemical modulation, is of particular interest [4]. This has been a promising approach because the biochemistry of 5-FU has been extensively studied for the past 30 years; therefore, effective biochemical strategies have been developed to exploit the activity of 5-FU and to attempt to overcome resistance to this agent [5]. These efforts have included combinations of 5-FU and hydroxyurea, methotrexate, cisplatin, thymidine, leucovorin and PALA (N-(phosphonacetyl)-L-(aspartate), respectively [6].

One interesting approach to modulation of 5-FU activity has been the use of the antiviral compound interferon (IFN). Interferons are natural agents to be investigated as modulators because of their wide variety of biological effects. The mechanisms of interaction between interferon-α (IFN-α) and 5-FU are unclear. Since the apparent improved response rates achieved with 5-FU plus IFN-α cannot be ascribed to combining 2 active agents, as IFN-α lacks single-agent activity in colorectal carcinoma, IFN may biochemically modulate the action of 5-FU as observed in preclinical models [7,8].

The pharmacokinetics of 5-FU may also be altered by IFN-α as a result of decreased 5-FU clearance from plasma [9]. An increase in manifested 5-FU toxic effects such as mucositis and diarrhoea as observed with this regimen supports this pharmacokinetic observation.

In renal cancer, where no effective chemotherapy exists, IFN-α as a single agent has shown beneficial effects with response rates of about 16%; the addition of the presumably inactive drug 5-FU may double response rates. In gastrointestinal cancer, 5-FU remains the most active single agent with response rates of about 20%; addition of the biological IFN-α may double response rates [10].

The incorporation of biological agents, often called "biological response modifiers", such as IFN-α into regimens with standard chemotherapeutic agents offers an important challenge since the assumptions for their use probably differ from those for chemotherapeutic agents. Their poorly understood mechanism of action, their relatively weak or absent cytotoxic activities, the wide range of biologically effective doses and the absence of a clear correlation between maximum tolerated dose and optimal therapeutic effect reflect factors impeding the development of rational strategies for incorporation of these compounds into clinical regimens [11-14]. Thus, it is far from clear what the optimal strategy for combining cytotoxic agents and biologicals might be and further strategies are required to elucidate the mechanism of action and interaction.

5-Fluorouracil

5-FU was synthesised as a uridine analogue, based on the observation that tumour cells preferentially take up uridine. The fluorine atom is placed on the pyrimidine ring in the position normally occupied by the methyl group, which distinguishes thymidine (Thd) from uridine.

The mechanism by which 5-FU exerts its antitumour effects is unknown, but at least 3 mechanisms have been postulated (Fig. 1). Firstly, 5-FU is converted to fluorodeoxyuridine monophosphate (FdUMP) through the pyrimidine salvage pathways; transfer of the ribose-phosphate group from phosphoribosylpyrophosphate (PRPP) effectively traps 5-FU within the cell. 5-FU is not affected by the multidrug transporter associated with chemotherapeutic drug resistance. Subsequently, FdUMP binds to the enzyme thymidylate synthetase (TS), forming a stable ternary covalent complex with 5,10-methylene tetrahydrofolate (5,10 MeTHF) and the enzyme. This complex has a half-life of 8 hours, effectively eliminating the enzyme from biochemical processes. As a result, the conversion of uridine metabolites to Thd is blocked. The intracellular pool of deoxythymidine triphosphate (dThdTP) necessary for DNA synthesis and repair is thus depleted [15].

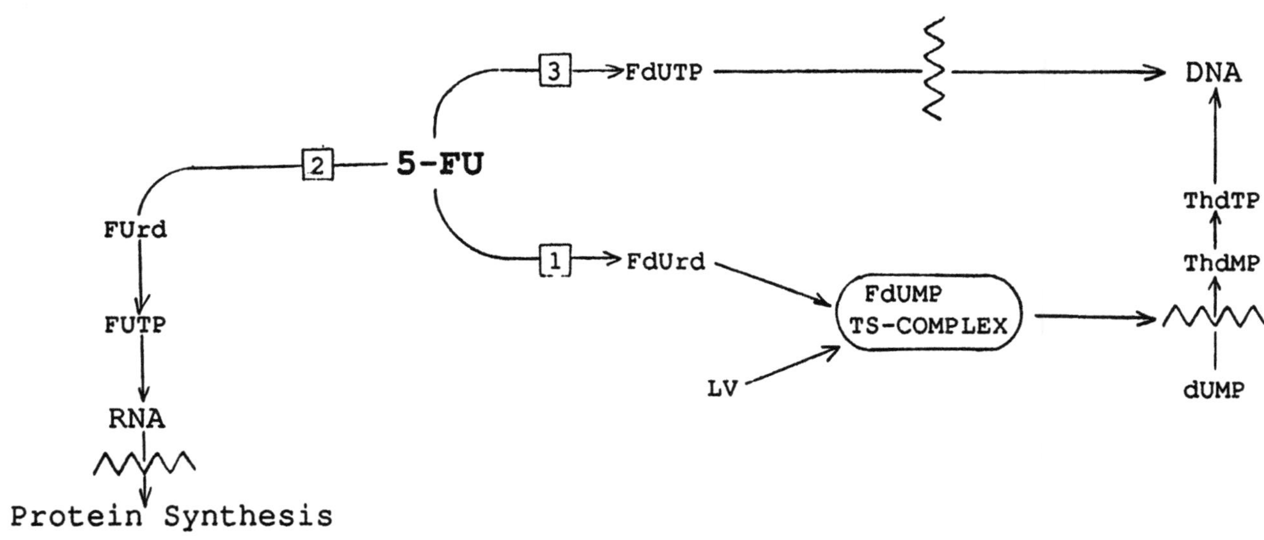

Fig 1. Influence of 5-FU on metabolic pathways:
1   5-FU inhibits DNA-synthesis by binding to the enzyme TS, forming a stable ternary complex with 5,10 MetHF and the enzyme. LV strengthens the 5-FU/TS-complex; FdUrd, 5-fluorodeoxyuridine; FdUMP, 5-fluorodeoxyuridinemonophosphate; TS, thymidilate synthetase; LV, leucovorin; dUMP, deoxyuridine; ThdMP, thymidinemonophosphate; ThdTP, thymidinetriphosphate;
2   5-FU inhibits protein synthesis via its incorporation into RNA as a false base; FUrd, 5-fluorouridine; FUTP, 5-fluorouridinetriphosphate;
3   Incorporation of FdUTP (resembling dTTP) into DNA results in DNA strand breakage and subsequent cell death; FdUTP, fluorodeoxyuridinetriphosphate; dTTP, deoxythymidinetriphosphate

Secondly, 5-FU may be converted directly into fluorouracil triphosphate (FUTP), resembling uridine triphosphate (UTP) required for RNA production. Incorporation of FUTP into RNA in the place of UTP results in malfunction of ribosomal RNA [16].

Thirdly, 5-FU may be converted to fluorodeoxy-uridine triphosphate (FdUTP), resembling deoxythymidine triphosphate (dThdTP) required for synthesis and repair of DNA; incorporation of FdUTP into DNA results in DNA strand breakage and subsequent cell death [17].

All the postulated mechanisms for 5-FU cytotoxicity require the conversion of 5-FU to its active antimetabolite.

Because the biochemistry of pyrimidine metabolism in humans is well established, and a number of drugs besides 5-FU affect these pathways, several attempts at biochemical modulation of pyrimidine biosynthesis have been made. The catabolic fate of 5-FU is the reduction of the pyrimidine ring by dihydropyrimidine dehydrogenase, an enzyme found mainly in liver and blood mononuclear cells.

In clinical trials, blockade of the enzyme resulted in persistent elevation of 5-FU levels and increased tumour kill, but also in increased toxicity. In an attempt to increase the incorporation of 5-FU into RNA, mentioned by some to be the major pathway of 5-FU action, exogenous Thd has been given. Thymidine would bypass the block in thymidylate synthetase induced by 5-FU, allowing DNA synthesis to proceed unhindered, while shunting 5-FU into RNA. This approach produced increased bone marrow toxicity without an increase in therapeutic advantage [18].

Clinical trials of 5-FU in combination with N-(phosphonoacetyl)-L-aspartate (PALA), which blocks the *de novo* synthesis of pyrimidines, were disappointing, but the drugs may not have been administered optimally [19]. Follow-up studies using PALA at a modulating dose, and 5-FU at clinically effective doses, resulted in 40% response rates in patients with colon cancer. Methotrexate, which antagonises dihydrofolate reductase, may act synergistically with 5-FU by increasing intracellular PRPP, but proper scheduling of drug delivery appears critical; optimal increases in PRPP occur at 24 hours after methotrexate administration [20].

Leucovorin has been used in an attempt to increase intracellular stores of 5,10 MeTHF, necessary for the formation of the ternary complex of FdUMP with TS. Reported response rates are in the range of 40% in previously treated patients, and may be higher in untreated patients with good performance status [21].

This therapy has the advantage of low toxicity, consisting mainly of diarrhoea; other less common toxicities include acute cerebellar syndrome and peripheral neuropathy [21]. *In vitro* studies, however, suggest that the most effective cell kill results from the combination of at least 3 agents affecting pyrimidine metabolism [22].

Interferon-α

IFN-α, originally described because of its ability to "interfere" with the *in vitro* cytopathic effects of the influenza virus, also produces antiproliferative effects in susceptible cells [23].

Like 5-FU, the mechanism of IFN's direct antitumour effect remains poorly understood.

IFN-α has a number of biochemical actions, many of which can be attributed to gene activation and the stimulation of the synthesis of several proteins of known and unknown functions like modulating cell surface antigens [24] or affecting several oncogenes [25].

One of the predominant cellular effects observed *in vitro* is the inhibition of cell cycle progression, with partial block in either the transition from G_0-G_1 to S, progression through S, or even generalised inhibition of cell cycle traverse [26].

IFN-α affects the metabolism of both polyamines and Thd [27,28] like the inhibition of the membrane transport of Thd [29], or its phosphorylation by Thd kinase (the rate-limiting step in Thd utilisation), and inhibits the conversion of uridine to Thd by salvage pathways [30]. The metabolism of other nucleotides, however, is not affected. It is still not completely clear whether the antitumour effects of IFN in humans are due to direct effects on the tumour, or indirect effects via the host.

Although IFN-α has shown therapeutic activity in several haematological malignancies, its role as a single agent in adenocarcinomas and other solid tumour entities has been disappointing. IFN-α as a single agent is inactive in colon cancer. Several trials have been published reporting only rare PRs. Gutterman and coworkers initially described 3 responses in 14 patients suffering from gastrointestinal cancer treated with natural (Cantell) IFN given on a daily basis [31]. Six subsequent trials using doses of 3 to 50 million units/m^2 on twice-a-week schedules or 3-times weekly obtained only 3 responses in 399 patients with gastrointestinal cancer.

In renal cell carcinoma, IFN-α yielded modest activity with overall response rates of 16% and response durations averaging 8 months. In a meta-analysis performed by Meadows and Ozer in more than 1,120 patients, daily IFN therapy seemed marginally superior to other schedules [32].

The optimal dose of IFN-α for renal cell carcinoma has not been defined yet, but is clearly not the maximal tolerated dose. It may easily be possible that very low doses of IFN, given on a daily schedule, are at least as effective as higher, less well tolerated doses [32].

Because of the relatively weak cytotoxic effects of IFN-α in solid tumours, it has been postulated that they may be best used in combination with other cytotoxic agents [33].

Combination of IFN-α with 5-FU

Studies evaluating the antitumour activity of IFN in combination with cytotoxic agents were begun shortly after the antitumour activity of IFN was recognised in experimental animal tumour models [36,37]. The earliest studies against murine leukaemias were largely empirical in design and were based on the assumption that cytotoxic agents were most useful for debulking large tumour volume and that the resultant microscopic residual disease would best be eradicated with IFN "immunotherapy" [14]. Anyhow, the therapeutic approach of combining IFN with 5-FU has been hampered by the lack of understanding of the optimal strategies for combining biologicals and cytotoxics. The use of biologicals, such as IFN-α, is complicated by their effects on the host immune system and on drug metabolising enzymes such as the cytochrome P-450 system. Some drug combinations appear additive but others may in fact represent synergistic combi-

nations, suggesting that IFN may alter the metabolism of cytotoxics in novel ways.

Antiproliferative Synergy of 5-FU and IFN-α in Model Systems

The rapidly enlarging literature on preclinical studies on the antineoplastic activity of the combination of IFN with various cytotoxic agents against experimental and human malignancies was reviewed by Wadler and coworkers [32].

Miyoshi and coworkers were the first to describe *in vitro* synergy between 5-FU and IFN-α [39]. They found that 5-FU added simultaneously with fibroblast IFN produced synergistic cytotoxicity for HeLa human cervical carcinoma [45,46], KMM-1 human myeloma cell line [39,47-49], and a transformed embryonic lung fibroblast cell line [39,47]. Additive effects could be observed in Raji human Burkitt's lymphoma cell line [39,47,49]. No additive effects were found in MCF-7 breast cancer or WI-38 normal lung fibroblast cell lines [39,47-49].

The use of a system such as a mouse xenograft model of human gastrointestinal cancer has the advantage that the effects of IFN on the host are eliminated, as human IFN is inactive in mouse models. In a renal subcapsular assay system, 1 colon cancer and 2 gastric cancer cell lines exhibited additive or synergistic cytotoxicity, whereas a second colon cancer cell line showed subtractive effects, suggesting heterogeneity of tumour response [38]. The combination of 5-FU with IFN-α was shown to inhibit the growth of human salivary gland adenocarcinoma and, with IFN-α/ß, to inhibit the growth of mouse adenocarcinoma [40].

Elias and Crissman evaluated the human myeloid leukaemia HL-60 and the murine MCA-38 adenocarcinoma cell line [40]. Partially purified murine IFN-α/ß did not enhance the effect of 5-FU alone in inhibiting MCA-38 growth in their investigations. IFN-γ did interact with 5-FU and resulted in synergistic inhibition.

The human myeloid cell line HL-60 was shown to be more effectively inhibited by the combination of 5-FU and recombinant IFN-α than was the murine adenocarcinoma line inhibited by the combination of 5-FU and IFN-α/ß [40]. With the sensitive HL-60 cell line, addition of IFN-α to 5-FU resulted in up to 90% growth inhibition at 5 days of culture. Addition of IFN-α

also lowered by 10-fold the concentration of 5-FU required for cell kill in this human promyelocytic cell line.

Namba and colleagues studied 6 human cancer cell lines and a normal human embryonic fibroblast line [41]. The cell lines were cultured under continuous exposure to 5-FU at concentrations up to 4 µg/ml and concentrations of purified human fibroblast IFN up to 2,000 IU/ml. In 5 of the 6 human tumour cell lines tested, a clear additive inhibition of cell proliferation or of colony formation was to be reported. IFN did not enhance the cytotoxic effect of the mammary carcinoma cell line MCF-7 [39,47-49] or the normal human fibroblast line WI-38 [39,47]. For several tumour cell lines, marginally effective concentrations of 5-FU were found, but by combining these concentrations with IFN, clear inhibition occurred.

Welander and coworkers investigated the human ovarian cell line BG-1 in an assay of colony formation during 7-21 days in culture. Recombinant IFN-α (concentration of 10,000 IU/ml) showed no additive effect when combined with 5-FU (concentration of 6 µg/ml) [42]. Wadler et al. tested 2 human colon cancer cell lines, SW-480 and HT-29 [44]. They were able to show that the addition of recombinant IFN-α could reduce the 50%-inhibitory dose of 5-FU (concentration capable of reducing the number of clonogenic cells by 50%) by about half. IFN-γ was not shown to be more active than IFN-α in their experiments, contrary to IFN-ß.

From the work of Meadows and coworkers with 5-FU and IFN it resulted that *in vitro* IFN-α altered the biochemistry of tumour cells resistant to its antiproliferative effects [43]. The treatment of the promyelocytic cell line HL-60 with IFN-α did not affect tumour cell growth or oncogene expression, whereas incorporation of radiolabelled Thd into DNA was inhibited by 49%.

Normally, incorporation of Thd serves as a marker for cell growth. IFN inhibits, however, the enzyme Thd kinase, which regulates conversion of Thd into a form suitable for DNA synthesis.

In the HL-60 cell line the cells were presumably able to overcome the block in Thd kinase through pyrimidine salvage pathways. This suggested that IFN could alter the biochemistry of a cell that was resistant to the growth inhibitory effects of IFN; moreover, an agent such as 5-FU, which affects pyrimidine biosynthesis at another point, might cooperate or synergise with IFN to inhibit tumour growth [40].

Taken together, these various reports indicated that IFN-α may interact with 5-FU to produce a modest antiproliferative response in some cell lines. However, it is not clear whether the modest level of inhibition sometimes seen in the idealised circumstances of a laboratory would translate to a striking response rate in the clinical setting.

Mechanisms of Interaction of 5-FU and IFN-α

The experiments described above cannot answer the question of the mechanism of interaction between IFN and the cytotoxic agent. It seems likely that the interactions observed were not solely the consequence of the combined effect of 2 cytoreductive agents, since the enhanced activity of the 5-FU/IFN combination was observed even in instances where IFN alone lacked activity, and IFN also failed to potentiate the 5-FU efficacy.

There are at least 4 better defined hypothetical modes of interaction of 5-FU and IFN:

(1) There may be a direct cytotoxic or antiproliferative interaction of IFN-α with 5-FU by an effect of IFN upon one of the pathways through which 5-FU affects DNA or RNA function;

(2) IFN-α may affect the pharmacokinetics or disposition of 5-FU in a manner that increases the concentration or the duration of exposure of 5-FU;

(3) IFN-α has a variety of actions of potential usefulness in the treatment of cancer including potent "immunomodulatory" effects. 5-FU may modulate these pathways affected by IFN through inhibition of protein synthesis or other effects;

(4) IFN-α may protect host tissues from the toxic effects of 5-FU via an indirect mechanism, allowing the use of higher doses of cytotoxic agents.

Some evidence exists for each of these hypothetical interactions:

ad 1: IFN might biochemically modulate the activity of anticancer drugs, e.g. by affecting critical target enzymes, repair mechanisms, or detoxification pathways within the tumour cell. Experiments using 5-fluorouracil analogues and those demonstrating partial reversal of inhibition by the addition of thymidine suggested that the interaction of 5-FU and IFN occurs through the target enzyme thymidylate synthetase, which acts in the pathway for the methylation of uridine. However, not all the effects of 5-FU were reversible by thymidine. These data would suggest that the enzyme thymidylate synthetase might be a target for the 5-FU/IFN interaction.

ad 2: The pharmacokinetics of 5-FU may also be altered by IFN-α as a result of decreased 5-FU clearance from plasma [55]. An increase in manifest 5-FU toxic effects such as mucositis and diarrhoea as observed with this regimen supports this pharmacokinetic observation.

Meadows and coworkers described alterations of pharmacokinetics in the disposition of 5-FU by IFN [56]. They measured 5-FU levels by high-pressure liquid chromatography in their patients. Patients were treated with 5-FU by continuous infusion for 24 hours, then given the first injection of IFN. In most patients, 5-FU was at or below the limits of detection (1 ng/ml) even after 24 hours of 5-FU infusion. However, after a single injection of IFN, the levels of 5-FU rose up to 16-fold within 1 hour and all patients investigated had detectable levels; this elevation in plasma 5-FU levels persisted 4 and 24 hours later. These data give additional support that altered 5-FU catabolism induced by IFN may be responsible, at least in part, for the synergism observed. Grem and colleagues have demonstrated that the clearance of 5-FU is altered in patients additionally receiving IFN-α and leucovorin [55]. Nevertheless it seems unlikely that the combination of 5-FU and IFN in clinical terms is no more than the functional equivalent of higher dose 5-FU. Moreover, the pharmacokinetic considerations would not explain the interaction and synergy observed *in vitro* and described above.

ad 3: IFN is known to have significant immunological effects when tested *in vitro* and when administered to patients. It has been shown to activate various classes of cytotoxic effector cells capable of killing tumour targets. This activation may occur *in vitro* as well as among the circulating cells harvested from patients after treatment with IFN [57,58]. IFN-α and IFN-γ increase the expression of various cell surface molecules including human leukocyte antigen (HLA) class I antigens [59]. This effect has also been observed in the circulating cells of patients treated with IFN and *in vitro* with regard to both lymphocytes and tumour targets [60]. Enhancement of the expression of HLA antigens may render tumour targets more susceptible to cytotoxic T cells. Enhancement of the expression of adhesion molecules may increase the binding efficiency of killer cells and targets. Targets incubated with IFN may become resistant to the cytotoxic action of natural killer cells and lymphokine-activated killer cells. The efficiency of this resistance seems to be directly correlated to the intensity of expression of HLA class I antigens, as has been shown in a variety of ways [61]. It has been observed that IFN could be administered to cancer patients in demonstrably immunomodulatory doses, yet for many common solid tumours clinical tumour response has been rarely observed. Thus, in both colon cancer and in melanoma, Neefe et al. demonstrated a significant increase in killer cell activity but very few clinical tumour responses [62]. A possible explanation for this disparity may be that the up-regulation of killer activity induced by IFN is coupled with a down-regulation of target sensitivity. Neefe and Glass recently investigated the effect of 5-FU upon the induction of tumour target resistance to IFN-α activated killers in the A-375 human melanoma cell line [63]. Recombinant IFN-α in concentrations from 100 to 10,000 IU/ml reproducibly diminished sensitivity of the target cells to lysis by IFN-α activated human peripheral blood lymphocytes as a source of killers by 50% to 90%. 5-FU alone had some tendency to enhance sensitivity of targets to killer cells without enhancing spontaneous death in the cytotoxicity assay. Targets treated with 5-FU before exposure to IFN failed to become resistant and, in some cases, showed to be substantially more sensitive to killer cells. The described 5-FU effect required relatively high doses of 5-FU (at least 3 µg/ml) and an exposure time of more than 1 hour for this cell line model. The effect of 5-FU upon IFN-induced HLA expression was investigated in the same cell line by Markovic and coworkers [64]. In parallel with the blockade of target cell resistance, 5-FU is capable of preventing the increased expression of HLA class I expression induced by IFN. Recombinant IFN-α increases the expression of HLA class I antigens. 5-FU has relatively little effect upon these antigens, but when given prior to IFN, 5-FU diminishes or even blocks the IFN effect.

In the model of resistance the effect of 5-FU is not blocked by performance of the experiment in the presence of 10 μmol/l thymidine. This observation indicates that the 5-FU effect is likely to be mediated through inhibition of RNA synthesis. This finding is also confirmed by the observation that 5-fluorouridine is highly effective in abrogating resistance whereas 5-fluorodeoxyuridine is reported to be ineffective [63]. PALA, an investigational inhibitor of protein synthesis in the uridine synthetic pathway, can also block the induction of resistance by IFN. It should be noted that the same concentrations of PALA and 5-FU found to block resistance induction by IFN have no effect upon the activation of natural killers by IFN. It has clearly been demonstrated by several investigations that 5-FU may influence the immunomodulatory actions of IFN. It has also been found that the activity of circulating killer cells may be increased by IFN. But evidence that these cells are capable of reaching patients' tumours in sufficiently large numbers, or that they have the potential of eradicating large numbers of cancer cells remains restricted to animal models [64,65].

ad 4: An alternative indirect mechanism for the interaction of IFN and 5-FU was reported by Stolfi et al. Partially purified or recombinant IFN-α was found to protect mice from the toxic effects of 5-FU [66,67]. This protection was manifested as decreases in leukopenia, body weight loss and mortality. The schedule of administration of the 2 agents, 5-FU followed by multiple injections of IFN, was similar to that used in many of the *in vivo* studies reported. Stolfi and coworkers suggested that the mechanism for the protective effect of IFN was the suppression of proliferation of the normal bone marrow cells of the host, rendering them less sensitive to the cytotoxic actions of 5-FU. Presumably the protective effect of IFN would allow higher doses of cytotoxic drugs to be used, thus increasing their antitumour activity [67].

Clinical Trials with 5-FU and IFN-α

Based on preclinical data suggesting synergy between 5-FU and IFN in colon cancer cell lines [68], Wadler and colleagues at the Albert Einstein Medical Center in New York initiated a clinical study in patients with advanced measurable colorectal cancer.

The enhancement of 5-FU cytotoxicity by IFN was both dose and schedule dependent. IFN was shown to be active added to or after 5-FU, but not before. Combination therapy with 5-FU and IFN, as originally designed, used doses and schedules of the 2 agents that were derived empirically. The dose and schedule of IFN used represent a conventional regimen that approaches the maximum tolerated dose [34,35].

Wadler gave the 5-FU by a loading dose schedule, followed by weekly bolus administration; IFN-α was administered 3 times a week.

The response rate to 5-FU plus IFN-α therapy in previously untreated patients reported by Wadler has been higher than 60%. Interestingly, no responses were observed in previously treated patients. However, toxicities seen in this therapeutic regimen included fatal diarrhoea in the setting of leukopenia. Myelosuppression and mucositis were also observed, along with fatigue, decrease in performance status, and CNS symptoms [34].

Other clinical trials with the combination have not been as successful as Wadler's, including a larger Eastern Cooperative Oncology Group-sponsored trial with response rates approaching 35% to 40%. These data again underlined the need to study the effects of schedule, route, and dose of the regimen in the clinical setting.

De Vecchis and coworkers reported on 12 patients suffering from advanced colorectal cancer treated with 5-FU plus IFN-α. The regimen was well tolerated with 2 responses observed [69].

Kreuser et al. reported only 1 partial response in 6 patients using a regimen of IFN-α2c, given twice weekly and 5-FU, administered at 10mg/kg i.v. continuously [70].

In a randomised phase II study performed by Clark et al. in advanced cancer, 29 patients were treated with 5-FU for 5 days, and with IFN-α given either at 20 x 10^6 IU/day i.v. for 5 days, or 5 x 10^6 U/day subcutaneously; only one patient responded objectively [71].

In a phase I trial, Meadows and colleagues treated 17 patients with advanced malignancy using a 5-FU/IFN regimen. 5-FU was given by continuous infusion at 300 mg/m^2 for 8 weeks, using the Lokich's regimen [72]. IFN-α was given in escalating doses to cohorts of 3 to 6 patients - 2, 3.5, 5, and 10 x 10^6 IU/m^2/day - beginning the day after 5-FU administration. The major dose-limiting toxicity was reported in mucositis, which occurred in 10 patients, requiring interruption of therapy and dose reduction of 5-FU. Other observed toxicities, like myelosuppression, alopecia, or rise in liver enzymes, were all mild. No patient developed neutropenic sepsis, and no toxic death occurred. Eight out of 17 patients developed rapid disease progression while on treatment, one patient withdrew after 1 week, and 8 patients completed at least one full course of planned therapy. The overall dose of 5-FU delivered in patients completing at least 14 days of therapy was 160 mg/m^2/day. IFN-α doses above 2 x 10^6 IU/m^2/day resulted in increased toxicity. They reported 2 complete responses, 2 partial responses, and 2 minor responses, all in patients with gastrointestinal cancer. Three out of the responding patients had previously failed to 5-FU-containing treatment schedules.

The data obtained in this trial led to the recommendation of exploring a dose of 250 mg/m^2/day of 5-FU, and 1 to 3 x 10^6 IU/day of IFN-α for further trials. In addition, patients should receive allopurinol mouthwash and prophylactic pyridoxine to prevent mucositis and hand-foot syndrome, respectively. The relatively mild toxicity observed, including minimal myelosuppression, suggested that other drugs that "modulate" pyrimidine biosynthesis, such as PALA or leukovorin, could be safely added to the regimen in future trials [73].

It is possible that the current randomised trials comparing 5-FU plus IFN-α to 5-FU plus folinic acid or 5-FU plus folinic acid plus IFN-α will contribute to define the activity, toxicity, and survival associated with these regimens.

In renal cell carcinoma, Sella and coworkers treated 49 patients using a 5-FU/IFN-α regimen. Patients received 5-FU, given by a 5-day infusion, and IFN-α given 3 times a week with or without mitomycin C. They reported an overall objective response rate of 35%. No unexpected toxicities occurred; mitomycin C could not contribute to the combination, other than toxicity [74], but further studies are needed to confirm this finding and to assess the role of mitomycin C.

Walther et al. initiated a pilot phase II study in advanced renal adenocarcinoma using continuous infusion 5-FU (300mg/m^2/day) and IFN-α (3.5 x 10^6 IU/m^2/day), in an outpatient setting presenting with about the same response rate as described above. Toxicities observed included mucositis and myelosuppression [75].

Although a limited number of patients have been treated in these studies so far, the results are noteworthy, as they represent the first attempts to improve on the responses seen with IFN-α alone.

Future strategies are required to investigate alternative schedules which reduce toxicity and further elucidate the mechanism of interaction.

Clinical Toxicities of the Combination of 5-FU and IFN-α

Combination treatment of 5-FU and IFN-α has resulted in a novel constellation of clinical toxicities including constitutional symptoms, gastrointestinal toxicities, myelosuppression and cerebral symptoms, as well as less common toxicities such as allergic reactions. It is interesting to note that the clinical spectrum of toxicities observed with this combination differs from those seen with either 5-FU or IFN-α alone [76].

Constitutional symptoms are the most common toxicity. During the first week of therapy, fevers and chills are frequent, but tend to abate by the second week. After the second week, many patients develop fatigue or malaise, which seem to be dose-related phenomena. Intermittent therapy results in lower fatigue levels than daily therapy. The level of fatigue is highly variable with some patients having none and some patients having persistent, incapacitating fatigue despite multiple dose reductions of IFN-α in the combination.

Gastrointestinal symptoms are the second most common group of symptoms observed with this treatment regimen [76]. Stomatitis in one of various forms seems to be predominantly associated with the continuous infusion phase of 5-FU treatment. The most common finding is pain with oral erythema. Less frequently, patients may present with either a sore throat or cracking at the corners of their lips without any oral cavity findings. Ulcers, either punctate or broad and shallow, are late findings consistent with manifest stomatitis.

The second most common sign of gastrointestinal inflammation is diarrhoea. Watery diarrhoea, defined as liquid, unformed stools, usually associated with tenesmus and cramping, may occur repeatedly during the course of a single day and has been frequently found to be a harbinger of myelosuppression. This may occur at any time during the course of treatment. As a serious complication, patients with watery diarrhoea while receiving 5-FU/IFN-α are reported to develop absolute neutropenia followed by life-threatening or fatal sepsis.

Allergic reactions are described in a minority of patients, mostly manifested as a rash that usually is erythematous and maculopapular. Painful erythema of the palms and soles is observed occasionally [76].

Management of patients receiving this regimen, which may be prolonged and intensive, requires close monitoring of these toxicities and careful evaluation of signs and symptoms.

Summary

The combination of 5-FU and IFN-α may represent a significant advance, heralding a successful mixture of cytotoxics and biologicals. However, many questions and problems remain unresolved.

With a few exceptions, enthusiastic reports of synergy between IFN and chemotherapeutic agents in preclinical systems cannot be translated yet into an important benefit in clinical trials. This may be due to the failure of the preclinical systems studied to accurately reflect the clinical setting, or to a failure to apply the preclinical findings to the clinic. For example, in combination with cytotoxic agents, sequence and duration of exposure to IFN may play as significant a role as dose and dose-intensity, and the maximum tolerated dose of IFN may not be the most biologically effective dose. The optimal dose and schedule of IFN still remain to be determined. There are suggestions that lower doses of IFN may be as effective as higher doses. It also remains to be seen whether daily IFN is better than a 3-times-weekly regimen, or whether continuous 5-FU infusion is better than loading-dose 5-FU.

What is the optimal time between 5-FU administration and IFN-α application?

The regimens used so far have been empirically derived, with no solid basis in an understanding of the cellular pathways in which the two agents may interact. The lack of understanding of the biochemical interactions of the combined agents has prohibited a rational approach to design of schedule, sequence, and dose relationship which allows translation of the positive preclinical data into effective clinical treatment regimens.

The proliferation of some human cell lines was shown to be significantly inhibited by 5-FU and IFN in clinically achievable concentrations. It has become evident from several investigations that a target for this action may be the enzyme thymidylate synthetase. This evidence is also founded on the demonstration that the presence of thymidine in the culture medium tends to block the synergistic effect and that the deoxyribosylated analogue of 5-FU is more effective than the RNA precursor.

It has also been demonstrated that IFN may affect the disposition of infused 5-FU in a manner that appears to be the functional equivalent of increasing the 5-FU dose.

A third possible interpretation of the interaction of 5-FU and IFN involves the phenomenon of IFN-induced resistance of tumour targets to lysis by IFN-α activated killer cells. 5-FU abrogates the resistance phenomen, possibly without affecting the activation of killer cells by IFN. This effect of 5-FU is not blocked by thymidine and is generated more readily by the ribosylated analogue of 5-FU than by the deoxyribosylated analogue. The effect of 5-FU in the resistance system may be mediated through modulation of expression of HLA class I antigens.

Furthermore, reports that IFN may protect host tissues from the toxic effects of 5-FU offer further opportunities to examine an indirect mechanism of interaction between IFN and cytotoxic agents.

The use of IFN-α in renal cell carcinoma underscores the differences between biologicals and cytotoxics with regard to dose, toxicity, and time to response. With cytotoxic chemotherapy, the drug dose is pushed to the maximum tolerated level; responses are usually dose related and occur with one to two treatment courses. The optimal dose of IFN-α in renal cell carcinoma still remains poorly defined. A dose-response correlation for tumour shrinkage has not yet been established. Further, the time to response to therapy may be prolonged up to 6 months as described in some studies. Toxicities, however, are dose related. Thus, it is important to design regimens that can be tolerated for many months.

Similar questions with regard to dose and schedule of both 5-FU and IFN-α have to be answered in renal cell carcinoma and in gastrointestinal cancer. In gastrointestinal cancer 5-FU is considered the active drug, and IFN-α is the modulator. In renal cell carcinoma, the reverse may be true, which would have implications for the dose and schedule of IFN.

Up to now, there is no information which predicts that 5-FU and IFN-α, combined in these doses and schedules, interact optimally. Broader knowledge of the mechanisms and of possible loci of interaction in cellular metabolism may lead to an optimally adapted regimen with a possible higher response rate. Encouraging preliminary results with the combination of IFN-α and 5-FU suggest that effective combinations are achievable, and that further studies are warranted.

REFERENCES

1   Davis M: Chemotherapy of large bowel cancer. Cancer 1982 (50,11):2638-2646
2   Wadler S, Green M, Magram J: The role of anthracyclines in the treatment of gastric cancer. Cancer Treat Rev 1988 (12,2):105-132
3   Wadler S, Lyver A, Wiernik PH: Clinical toxicities of the combination of 5-fluorouracil and recombinant interferon-alfa-2a: an unusal toxicity profile. ONF 1989 (16,6Supplement):12-15
4   Leyland-Jones B, O'Dwywer PJ: Biochemical modulation: Application of laboratory models to the clinic. Cancer Treat Rep 1986 (70,1):219-229
5   Weber G: Biochemical strategy of cancer cells and the design of chemotherapy. GH Clowes Memorial Lecture. Cancer Res 1983 43(8):3466-3492
6   Santo DR, McHenry CS, Sommer H: Mechanism of interaction of thymidylate synthetase with 5-fluorodeoxyuridylate. Biochemistry 1974 (13):471-480
7   Elias L, Crissman HA: Interferon effects upon the adenocarcinoma 38 and HL-60 cell lines: Antiproliferative responses and synergistic interactions with haloginate pyrimidine antimetabolites. Cancer Res 1988 (48):4868-4873
8   Elias L, Sandoval JM: Interferon effects · upon fluorouracil metabolism by HL-60 cells. Biophysics Res Commun 1989 (163): 867-874
9   Grem JL, Allegra CJ, McAtee N, et al: Phase I study of interferon alfa-2a, 5-fluorouracil and high-dose leucovorin in metastatic gastrointestinal carcinoma. Proc Am Soc Clin Oncol 1990 (9):70
10 Meadows L, Ozer H: Treatment of gastrointestinal and renal adenocarcinomas with interferon-alpha. Biotherapy 1992 (4,3): 179-187
11 Dinarello CA, Mier JW: Lymphokines. N Engl J Med 1987 (317): 940-945
12 Pfeffer LM: Cellular effects of interferons. In: Pfeffer LM (ed) Mechanism of Interferon Actions. Vol. 2, CRC Press, Inc, Boca Raton, FL 1987 pp 2-18
13 Krown SE: Interferons in malignancy: biological products or biological response modifiers? JNCI 1988 (80): 306-309
14 Gutterman JU, Hersh EM: Immunotherapy. In: Holland JF, E Frei (eds) Cancer Medicine. Lea and Febinger, Philadelphia 1982 pp 1100-1132
15 Heidelberger C: Cancer chemotherapy with purine and pyrimidine analogues. Annual Rev Pharmacol 1967 (7):101-124
16 Chaudhuri NK, Mukherjee KL, Heidelberger C: Studies on fluorinated pyrimidines VII-The degradative pathway. Biochem Pharmacol 1958 (1):328-341
17 Kufe DW, Major PP, Egan EM, et al: 5-Fluoro-2'deoxyuridine incorporation into L1210 DNA. J Biol Chem 1981 (256):8885-8888
18 Au JL, Rustum Y, Ledesma EJ, et al: Clinical pharmacological studies of concurrent infusion of 5-fluorouracil and thymidine in treatment of colorectal carcinomas. Cancer Res 1982 (42):2930-2937
19 O'Connell MJ, Powis G, Rubin J, et al: Pilot study of PALA and 5-FU in patients with advanced cancer. Cancer Treat Rep 1982 (66):77-80
20 Benz C, Cadman E: Modulation of 5-fluorouracil metabolism and cytotoxicity by antimetabolite pretreatment in human colorectal adenocarcinoma HCT-8. Cancer Res 1981 (41):994-999
21 Madajewicz S, Petrelli N, Rustum Y, et al: Phase I-II trial of high-dose calcium leucovorin and 5-fluorouracil in advanced colorectal cancer. Cancer Res 1984 (44):4667-4669
22 Martin DS: Purine and pyrimidine biochemistry and some relevant clinical and preclinical cancer chemotherapy research. In: Powis G, Prough RA (eds) Metabolism and Action of Anti-Cancer Drugs. PA Taylor and Francis, Philadelphia 1987 pp 91-140
23 Gresser I, Brouty-Boye D, Thomas MT, et al: Interferon and cell division. I. Inhibition of the multiplication of mouse leukemia L1210 cells in vitro by interferon preparations. Proc Natl Acad Sci USA 1970 (66):1052-1058
24 Ruiter DJ, Brocker EB, Ferrone S: Expression and susceptibility to modulation by interferons of HLA class I and II antigens on melanoma cells. J Immunogenet 1986 (13): 229-234
25 Jonak GJ, Knight E Jr: Selective reduction of c-myc mRNA in Daudi cells by human beta interferon. Proc Natl Acad Sci USA 1984 (81):1747-1750
26 Tamm I, Jasny BR, Pfeffer LM: Antiproliferative action of interferons. In: Pfeffer LM (ed) Mechanism of Interferon Actions. Vol. 2, CRC Press, Inc., Boca Raton, FL 1987 pp 25-28
27 Watanabe T, Sherman M, Shafman T, et al: Effects of ornithine decarboxylase inhibition on c-myc 10 expression during murine erythroleukemia cell proliferation and differentiation. J Cell Physiol 1986 (127):480-484
28 Pfeffer LM, Tamm I: Interferon inhibition of thymidine incorporation into DNA through effects on thymidine transport and uptake. J Cell Physiol 1984 (121):431-436
29 Gewert DR, Moore G, Clemens MJ: Inhibition of cell division by interferons: The relationship between changes in utilization of thymidine for DNA synthesis and control of proliferation in Daudi cells. Biochem J 1983 (214):983-990
30 Barankiewicz J, Daplinsky C, Cohen A: Modification of ribonucleotide and deoxyribonucleotide metabolism in interferon-treated human lymphoblastoid B-cells. J Interferon 1986 (6):177-727
31 Gutterman J, Quesada J: Clinical investigation of partially pure and recombinant DNA derived leukocyte interferon in human cancer. Tex Rep Biol Med 1982 (41):626-633
32 Wadler S, Schwartz EL: Antineoplastic activity of the combination of interferon and cytotoxic agents against experimental and human malignancies: A review. Cancer Res 1990 (50):3473-3486
33 Lin SI, Kikuchi T, Pledger WJ, Tamm I: Interferons inhibits the establishment of competence in G_0/S phase transition. Science (Wash. DC) 1986 (233):356-359
34 Wadler S, Schwartz EL, Goldman M, et al: Fluorouracil and recombinant alfa-2a-interferon: An

active regimen against advanced colon cancer. J Clin Oncol 1989 (7):1769-1775

35 Pazdur R, Ajani JA, Patt YZ, et al: Phase II study of fluorouracil and recombinant interferon alfa-2a in previously untreated advanced colorectal carcinoma. J Clin Oncol 1990 (8):2027-2031

36 Gresser I, Bourali C, Levy JP, Brouty-Boye D, Thomas MT: Increased survival in mice inoculated with tumor cells and treated with interferon preparations. Proc Natl Acad Sci USA 1969 (63):51-57

37 Gresser I, Bouraly C: Antitumor effects of interferon preparations in mice. JNCI 1970 (45):365-376

38 Nishiyama M, Takagami S, Kirihara Y, et al: Combined effects of interferon alpha-A/D with fluoropyrimidine derivates in the subrenal capsule assay. Gan To Kagaku Ryoho 1988 (15):2285-2290

39 Miyoshi T, Ogawa S, Kanamori T, Nobuhara M, Namba M: Interferon potentiates cytotoxic effects of 5-fluorouracil on cell proliferation of established human cell lines originating from neoplastic tissues. Cancer Lett 1983 (17):239-247

40 Elias I, Crissman HA: Interferon effects upon the adenocarcinoma 38 and HL-60 cell lines: antiproliferative responses and synergistic interactions with halogenated pyrimidine antimetabolites. Cancer Res 1988 (48):4868-4873

41 Namba M, Myoshi T, Kanamori T, et al: Combined effects of 5-fluorouracil and interferon on proliferation of human neoplastic cells in culture. Gann 1982 (73): 819-824

42 Welander CE, Morgan TM, Homesley HD, Trotta PP, Spiegel RJ: Combined recombinant human interferon-α2 and cytotoxic agents studied in a clonogenic assay. Int J Cancer 1985 (35):721-729

43 Meadows LM, George DJ, Kaufman RE: Dissociation of thymidine incorporation and transferrin receptor expression from cell growth and c-myc expression in alpha-interferon treated cells. J Biol Response Mod 1990 (9):212-220

44 Wadler S, Schwartz EL, Wersto R, Thompson D, Wiernik PH: Interferon (IFN) modulates the activity of 5-fluorouracil against two human colon cancer cell lines. Proc Am Assoc Cancer Res 1989 (30):569

45 Namba M, Yamamoto S, Tanaka H, Kanamori T, Nobuhara M, Kimoto T: In vitro and in vivo studies on potentiation of cytotoxic effects of anticancer drugs or cobalt 60 y-ray by interferon on human neoplastic cells. Cancer 1984 (54):2262-2267

46 Yamamoto S, Tanaka H, Kanamori T, Nobuhara M, Namba M: In vitro studies on potentiation of cytotoxic effects of anticancer drugs by interferon on a human neoplastic cell line (HeLa). Cancer Lett 1983 (20):131-138

47 Namba M, Miyoshi T, Kanamori T, Nobuhara M, Kimoto T, Ogawa S: Combined effects of 5-fluorouracil and interferon on proliferation of human neoplastic cells in culture. Gann 1982 (73):819-824

48 Miyoshi T, Ogawa S, Kanamori T, Nobuhara M, Kimoto T, Namba M: Combined effects of 5-fluorouracil and interferon on proliferation of human neoplastic cells in culture. Gan to Kagaku Ryoho 1982 (9):2207-2211

49 Namba M, Yamamoto S, Tanaka H, Kanamori T, Nobuhara M, Kimoto T: Potentiation of the cytotoxic effects of various anticancer drugs by interferon on human neoplastic cells (HeLa) in culture. Jpn J Cancer Chemother 1983 (10):1308-1312

50 Von Hoff DD, Huong AM, Collins J: Effect of recombinant ß interferon (rIFN-b$_{ser}$) on primary human tumor colony forming units. J Interferon Res 1988 (8):813-820

51 Marquet RL, Jeekel J: Combined effect of 5-fluorouracil and interferon on experimental liver metastases of rat colon carcinoma. J Cancer Res Clin Oncol 1985 (109):156-158

52 Le J, Yip YK, Vilcek J: Cytolytic activity of interferon-α and its synergism with 5-fluorouracil. Int J Cancer 1984: 495-500

53 Gohji K, Maeda S, Sugiyama T, Ishigami J, Kamidono S: Enhanced inhibition of anticancer drugs by human recombinant γ-interferon for human renal cell carcinoma in vitro. J Urol 1987 (137):539-543

54 Morikawa K, Fan D, Denkins YM, Levin B, Gutterman JU, et al: Mechanisms of combined effects of γ-interferon and 5-fluorouracil on human colon cancers implanted into nude mice. Cancer Res 1989 (49):799-805

55 Grem J, Allegra CJ, McAtee N, et al: Phase I study of interferon-alpha-2a, 5-fluorouracil and high dose leucovorin in metastatic gastrointestinal cancer. Proc Am Soc Clin Oncol 1990 (9):272

56 Lindley C, Bernard S, Gavigan M, et al: Interferon-alpha increases 5-fluorouracil levels 16-fold within 1 hour: Results of a Phase I study. J Interferon Res 1990 (10,II): 9-5

57 Einhorn S, Blomgren H, Strander H: Interferon and spontaneous cytotoxicity in man. Acta Med Scand 1978 (204):477-483

58 Huddlestone JR, Merigan TC, Oldstone MBA: Induction and kinetics of natural killers in humans following interferon therapy. Nature 1979 (282):417-419

59 Giacomini P, Aguzzi A, Pestka S, et al: Modulation by recombinant DNA leukocyte (ß) and fibroblast (B) interferons of the expression and shedding of HLA and tumor associated antigens by melanoma cells. J Immunol 1984 (133):1649-1655

60 Giacomini P, Fraioli R, Calabro AM, et al: Class I major histocompatibility complex enhancement by recombinant leukocyte interferon in the peripheral blood mononuclear cells and plasma of melanoma patients. Cancer Res 1991 (51) 652-656

61 Storkus WJ, Hovell DN, Salter RD, et al: NK susceptibility varies inversely with target cell class I HLA antigen expression. J Immunol 1987 (138):1657-1659

62 Neefe JR, Phillips EA, Treat J: Augmentation of natural immunity and correlation with tumor response in melanoma patients treated with human lymphoblastoid interferon. Diagn Immunol 1986 (4):299-305

63 Neefe JR, Glass J: Abrogation by 5-fluorouracil of interferon-induced resistance of a melanoma cell line to activated killers. Cancer Res 1991 (51):3159-3163

64 Markovic SN, Murasko DM: Role of natural killer and T cells in interferon induced inhibition of sponta-

neous metastases of the B16FIOL murine melanoma. Cancer Res 1991 (51):1124

65 Morikawa K, Fau D, Denkins YM, et al: Mechanism of combined effects of α-interferon and 5-fluorouracil on human colon cancers implanted into nude mice. Cancer Res 1989 (49):799

66 Stolfi RL, Martin DS, Sawyer RC, Spiegelman S: Modulation of 5-fluorouracil-induced toxicity in mice with interferon or with interferon inducer, polyinosinic-polycytidylic acid. Cancer Res 1983 (43):561-566

67 Stolfi RL, Martin DS: Modulation of chemotherapeutic drug activity with polyribonucleotides or with interferon. J Biol Response Modif 1985 (4):634-639

68 Wadler S, Schwartz EL, Goldman M, Lyver A, Rader M, Zimmerman M, Itri L, Weinberg V, Wiernik PH: Fluorouracil and recombinant Alfa-2a-Interferon: An active regimen against advanced colorectal carcinoma. J Clin Oncol 1989 (7):1769-1775

69 De Vecchis L, Nunziata C, Ricci R, et al: Immunochemotherpy of advanced colorectal cancer with alpha-interferon and 5-fluorouracil. I. Toxicological studies. J Chemother 1989 (1):128-135

70 Kreuser ED, Porzsolt F, Digel W, et al: Interferon alpha-2c in combination with 5-fluorouracil for refractory colorectal carcinoma. Fifth National Cancer Institute-European Organisation for Research and Treatment of Cancer Symposium on New Drugs in Cancer Therapy, October 22-24, 1986, Amsterdam, The Netherlands

71 Clark RI, Slevin ML, Reznek RH, et al: Two randomized phase II trials of intermittent intravenous versus subcutaneous alpha-2-interferon alone (trial 1) and in combination with 5-fluorouracil (trial 2) in advanced cancer. Int J Color Dis 1987 (2):262-269

72 Lokich J, Gillings D, Gallo J, et al: Bolus versus infusion 5-fluorouracil for treatment of advanced adenocarcinoma. Proc Am Soc Clin Oncol 1986 (5):83

73 Meadows LM, Bernard S, Gavigan M, et al: Preclinical and clinical studies with 5-fluorouracil and interferon-alpha. Proc Am Soc Clin Oncol 1990 (9):465

74 Sella A, Logothetis A, Dexeus F, et al: Increased response rate with the combination of chemotherapy 5-fluorouracil and mitomycin C with interferon (Roferon) in patients with metastatic renal cell carcinoma. J Urol 1992 147(3):573-577

75 Walther PJ, Meadows L, Walker RA: Treatment of advanced renal adenocarcinoma with alpha-2b-interferon and continuous infusion 5-fluorouracil in an outpatient setting; a phase II feasibility pilot study. Submitted to Am Urol Ass

76 Wadler S, Lyver A, Wiernik PH: Clinical Toxicities of the Combination of 5-Fluorouracil and Recombinant Interferon-Alfa-2a. An Unusual Toxicity Profile. Oncology Nursing Forum,1989 (Supplement to Vol 16, 6):12-15

Malignant Melanoma

G. Parmiani and A. Mazzocchi

Division of Experimental Oncology D, Istituto Nazionale Tumori, Via Venezian 1, 20133 Milan, Italy

Due to its steady increase in recent years, malignant melanoma is becoming a much dreaded disease in many Western countries. This phenomenon has been attributed to a more frequent and intense exposure to the sun for recreational purposes as a consequence of the healthy socio-economic level of most of the Western populations.

Early diagnosis can improve the clinical course of the disease thanks to surgery, which is curative in most of the primary tumour lesions with a thickness of less than 0.8-1.0 mm. However, once a recurrence or a metastatic event takes place, the chances of eradicating the disease are small. Melanoma has indeed proved to be resistant to most of the standard anticancer therapies such as cytotoxic agents and radiation [1]. Resistance occurs despite many experimental and clinical observations which indicate melanoma cells as potentially immunogenic to the host, and describe lymphocyte infiltration of the neoplastic lesions [2]. Many explanations have been proposed for the resistance of melanoma cells to therapeutic agents but none of them appears satisfactory. It is, therefore, necessary to study the biology of this event further in order to have a better understanding of melanoma cells and use this knowledge to design more effective clinical trials. In particular, the susceptibility of melanoma patients to interferon alpha (IFN-α), although limited, suggests that the biological effect that this cytokine can exert both on melanoma cells and on the host should be investigated further. Attention should be paid not only to the immune system, but also to the other less well-known activities of IFN, such as its ability to modulate cellular enzymes and/or expression of growth factor receptors. Such pleiotropic effects make it difficult to elucidate the detailed mechanism(s) of antitumour activity *in vivo* of IFN.

This review will focus mainly on the preclinical effects of IFNs, including IFN-α and -γ, and before summarising the available clinical data, we shall also consider some scarcely explored areas, such as the non-immunological effects of IFN treatment.

Preclinical Aspects

During the last few years, many studies have been performed to characterise the antitumoural properties of IFNs. This family of molecules can exert direct antiproliferative activity on neoplastic cells, stimulate a variety of immune effector cells potentially able to destroy tumour cells, and modulate molecules relevant in the interaction of tumour cells with the immune system. Tumour cell lines of different histotypes have been examined, but melanoma has been the tumour most frequently investigated as a target of the biological effects of IFNs. This is due to the low responsiveness of melanoma patients to conventional anticancer therapies and to the reported host-immune system playing an important role in the outcome of the disease [2].

Antiproliferative Effect

All types of IFN (α, β, γ) can have an antiproliferative effect on *in vitro* melanoma cell lines. It has been shown that IFN-β exhibits the most effective antiproliferative activity, compared to IFN-α and IFN-γ [3]. More recently, this observation has been confirmed and extended by other investigators, who reported that IFN-β is more potent than IFN-α and IFN-γ when tested on 8 *in vitro* melanoma lines and on one melanoma cell line grown as a xenograft in nude mice [4]. However, it should be noted that the antiproliferative activity of IFN-γ *in vitro* is not necessarily predictive of *in vivo* efficacy. In fact, it has been reported that only one out of three human melanomas, susceptible to IFN-γ *in vitro* and adapted to grow as a xenograft in nude mice, was sensitive to the antitumour effect of IFN-γ *in vivo* [5]. The mechanism(s) by which IFNs exert their antiproliferative/cytotoxic effects on tumour cells *in vitro* is still not clear, nor is the reason why some cells are more or less sensitive or resistant to it. In some reports, the different susceptibility among cell lines to IFNs is associated with the binding affinity of IFN receptors. From the work described above [4], it seems that receptor affinities might, in part, determine the relative efficacy of the different IFNs, since the order of binding affinity correlates with the order of antiproliferative activity. Other investigators, however, [6] have reported that two cell lines, one resistant to IFN-γ and the other to IFN-α and IFN-β, possessed an equal number of functional receptors of similar binding affinity. Studies concerning the mechanisms of sensitivity, or resistance, or of different susceptibility to IFNs have therefore provided conflicting results. The heterogeneity observed with regard to the antiproliferative effect of IFNs among different melanoma lines, has also been seen in different tumour clones isolated from the same lesion. As shown by Mortarini et al. [7], 4 out of 13 clones isolated from a melanoma metastasis appeared resistant after treatment with 500 IU/ml of IFN-γ. However, in many experimental situations the combination of IFNs with other cytokines may allow the heterogeneity of response to the antiproliferative effects of IFNs, when used alone, to be overcome. The combination of IFN-α or γ with TNF-α has proved to be the most effective. The synergy seen with these two cytokines may be explained on the basis of evidence that IFN-γ can induce or augment the receptors for TNF-α, thus increasing the ability of the cells to respond to the latter cytokine [8].

Modulation of Cell Surface Molecules

Another effect of IFNs to be taken into consideration is their ability to modulate various cell surface molecules relevant to the interaction of tumour cells with the host immune system and to the metastatic process. The first to be analysed for susceptibility by IFN modulation were the major histocompatibility complex (MHC) encoded antigens or HLA. They could play a central role in the patient's immune response to tumours, since MHC molecules expressed on the tumour cell surface may present tumour-specific peptide antigens to T cells and, therefore, favour the generation of a cell-mediated antitumour immune response. Accordingly, it has been reported that tumour progression may correlate with a reduction in MHC-class I antigen expression in experimental models [9]. In human melanoma, an association between low level of MHC class I or high level of class II antigens with metastasis has been described [10]. The expression of MHC class II molecules (HLA-DR) in histological sections of primary melanomas is reportedly associated with a high risk of metastasis [11]. Although most of the melanoma cell lines express class I HLA-A, B and C, only 50-60% of them express HLA-DR [12,13]. The expression of MHC class II molecules on melanoma cells is of particular interest since the normal counterpart (melanocytes) lacks the constitutive expression of these antigens. All types of IFNs are able to upregulate the expression of MHC class I molecules, although the enhancing effect of IFN-γ is more marked than that of IFN-α and IFN-β [14,15]. With regard to class II HLA, IFN-γ is able to induce/enhance the expression of HLA-DR in almost all melanomas examined by different researchers, whereas IFN-α and IFN-β are less effective [12,15]. The mechanisms involved in the differential regulation by the three IFNs of class II HLA genes in most melanoma cell lines remain to be determined. However, it has been demonstrated that the increase in cell surface expression of MHC molecules by IFNs is due to an increased level of mRNA, thus suggesting that IFNs act at a gene level as enhancers of transcription [16].

Another class of molecules susceptible to IFN modulation is that of the adhesion molecules.

Their expression on tumour cells is crucial, both for the MHC-restricted cytotoxicity mediated by T cells and for non-specific lysis by NK and LAK cells [17-19]. In particular, two different pathways have been documented in the lymphocyte-tumour cell binding; a) LFA-1 (on lymphocytes)/ICAM-1 (on target cells) and b) CD2 (on lymphocytes)/LFA-3 (on target cells). All melanoma cell lines express ICAM-1 and LFA-3 but there are differences, mainly for ICAM-1, in the fluorescence intensity that is related to cell surface density of the molecule. Modulation experiments performed *in vitro* have indicated that ICAM-1 and, to a lesser extent, LFA-3 may be upregulated by IFN-γ [3,7]. Such differences in expression may result in differences in the *in vitro* invasion [19] and the *in vivo* progression of melanoma cells [20,21], with metastatic melanoma lesions expressing higher levels of ICAM-1 compared to primary tumours [20,21]. It is important to have a better understanding of the role of molecules such as HLA-DR and ICAM-1 in tumour-host interaction in order to evaluate whether their increased expression, which can be induced by IFN-γ treatment, may be of benefit to patients, and whether *in vivo* changes in the expression of MHC antigens, or adhesion molecules, may have any predictive value on the outcome of the disease.

Another family of molecules expressed on melanoma cells and relevant in the biology of metastasis are integrins. These glycoproteins play an important role in the interaction of neoplastic cells with proteins of the extracellular matrix and, therefore, in a critical event of the multistep metastatic process. It has been shown [22] that exposure of a melanoma cell line to IFN-γ can down-modulate the expression of some integrins. Other investigators [23] have reported that IFN treatment of melanoma cells does not change the integrin profile. At the moment too few melanoma lines have been examined for integrin modulation after IFN exposure to draw clear conclusions.

Melanoma-associated antigens (MAAs) represent a class of molecules expressed on the majority of melanoma lines and investigated for susceptibility to modulation by IFNs. Some authors have shown that IFN-α and γ are able to increase the expression of some but not other MAAs [12,24]. More recently, it has been reported that a high molecular weight MAA is not susceptible to modulation by α, β, γ IFNs [14]; other investigators have demonstrated that MAA may be down-modulated after IFN treatment [7,25]. These data suggest that the effects of IFNs on MAAs are heterogeneous and depend on experimental conditions, on the melanoma cell lines used and, in particular, on the antigen analysed. An interesting application *in vivo* of this biological effect of IFN has been reported by Rosenblum and co-workers [26] who observed in melanoma patients an increased distribution to tumour tissue of an 111In-labelled MAb to MAA p97 following treatment with IFN-α. This means a significant difference in the number of metastases identified in the patients previously treated with IFN compared to those who received only MAb.

Interestingly, it has been reported that IFNs-α, β, and γ stimulate the expression of melanocyte-stimulating hormone receptor on the surface of melanoma cells. This can be one of the mechanisms by which IFNs exert their antitumour activity, since differentiation of melanoma cells can reduce their malignancy [27]. It is known that melanoma cells can express receptors for several cytokines and growth factors [28], and that melanoma can produce a variety of such cytokines and growth factors [29], thus potentially causing autostimulation through the formation of autocrine loops [30].

It seems, therefore, of paramount importance to examine whether IFN-α or γ can interrupt or modify such loops, thereby providing a therapeutic effect, as reported in hairy cell leukaemia [31] and most likely in kidney cancer [32].

A summary of the modulation of cell surface molecules by IFNs is given in Table 1. On the whole, as far as the modulation of surface

Table 1 : *In vitro* modulation of cell surface molecules on melanoma cells by IFNs

Molecule	Modulation by IFN		
	-α	-β	-γ
MHC class I	>	>	>
MHC class II	≥	≥	>
ICAM-1	≥	≥	>
β1 integrins	ND	ND	≥
MAA	≥ or <[1]	≥ or <	≥ or <
MSH-R	>	>	>

1 depending on the MAA evaluated; > significantly up-regulated; ≥ weakly or not up-regulated; < significantly down-regulated

molecules by IFNs is concerned, it is important to underline that heterogeneity exists in the susceptibility to IFN-γ treatment, not only among different melanoma cell lines, but also within the same lesion [7,33]. Moreover, treatment with IFN-γ often causes the shedding of surface molecules, such as HLA, MAAs [34], and ICAM-1 [35]. The latter phenomenon might be relevant *in vivo* since Becker and co-workers [35] have reported that the non-MHC restricted cytotoxicity can be inhibited by soluble ICAM-1 and by melanoma cell culture supernatants containing shed ICAM-1. This mechanism could be implemented by melanoma cells to escape the host's immune reactions. Recent data on the correlation between the presence of soluble ICAM-1 in the sera of patients with melanoma and a bad prognosis corroborate such an idea [36].

Modulation of the Immune System

Besides the biological effects described above, IFNs can activate cells of the immune system. Many reports indicate that IFNs modulate NK and macrophage activity [37-39] and IFN-γ is also involved in the T-cell mediated immune response [40]. The fact that IFNs can affect both tumour and the immune cells makes the understanding of the mechanisms involved in the *in vivo* antitumour activity complex.
It has been shown [38] that pretreatment of mice with 4 daily injections of IFN-α resulted in a marked inhibition of the number of pulmonary metastases of B16 melanoma and an increase in NK activity. When treatment was extended to 10 daily injections prior to melanoma cell inoculation, either the inhibitory effect on metastases or the rise in NK activity were less evident. The role of host immune response in the antitumour activity of IFNs has been confirmed by other studies [41]. Two sublines of melanoma B16, one resistant and the other sensitive to the direct antiproliferative effect of IFN-α *in vitro* when inoculated in mice, responded in the same way to IFN treatment, thus suggesting that IFN may exert its antitumour activity *in vivo* through the host defense system. The mechanisms by which IFNs mediate antitumour activity *in vivo* may be different according to the type of IFN used. It has

been shown that IFN-γ may be more effective than IFN-β in the growth inhibition of s.c. implanted tumour and in the inhibition of experimental pulmonary metastases, and that IFN-γ stimulates mostly macrophages, whereas IFN-β increases NK activity [42]. However, pretreatment of melanoma B16 *in vitro* with IFN-γ has provided different results. In fact, the treatment of melanoma cells with IFN-γ increases the capacity of the tumour to develop lung metastases [43]. This effect is associated with an enhancement of MHC class I antigen expression and with a reduction of NK-cell susceptibility. The same treatment with IFN-α and β has no effect on metastatic ability. On the basis of these data it can be concluded that IFN, by increasing the MHC antigen expression on tumour cells, causes a reduction of NK cell susceptibility and, therefore, facilitates the metastatic spread. An interesting study [44] has investigated the impact of IFN-γ treatment of melanoma cells on non-specific and specific immune response. IFN-treated tumour cells showed a reduced susceptibility to NK activity, but an increased immunogenicity; both effects might be due to an enhancement of MHC antigen expression. From a clinical point of view, the reduced susceptibility to NK and macrophage by melanoma cells may be overcome by a more effective T-cell response. After the most recent studies on the effect of MHC class I antigen expression on the NK susceptibility [45], such apparent discrepancy can be reconciled, although we still do not know which of these effects may prevail in melanoma patients receiving IFNs.
Several groups have investigated the possible relationship between NK activity, which is thought to be boosted by *in vivo* treatment with IFN-α, and prognosis. Despite some correlation reported with head and neck tumours [46], several studies in melanoma failed to reveal a significant association between NK activity in the peripheral blood and clinical outcome [47]. No relevant studies have been carried out on the T-cell response of patients during IFN therapy of melanoma. Given the complexity and cost of these investigations, it can be concluded that only a collaborative effort of different groups can provide relevant data on this issue, considering also that NK activity of peripheral blood lymphocytes may be less relevant than that of

Table 2. Antitumour activity of IFN-α2 as a single agent in metastatic melanoma [49]

IFN	Dose/schedule MU/m^2	N. evaluable patients	CR/PR	Response rate
α2a	3-50, 3-7 d/w i.m.	189	9/15	12.6
α2b	10-100, 3-7 d/w i.m., s.c., i.v.	106	9/15	22.6
Total		295	18/30	16.2

similar lymphocytes acting at the site of tumour growth.

New strategies that could be useful in order to evaluate the antitumour effects of IFNs are now in progress. In fact, the possibility of obtaining melanoma cells capable of producing IFN constitutively, after retrovirus-mediated gene transfer [48], will be an effective means for investigating in a more comprehensive way the role of IFN in host-tumour interaction.

Clinical Aspects

It would be impossible to summarise all the studies carried out with IFNs in melanoma patients during the last few years. Excellent reviews and books have been published on this subject [49]. We will only give a schematic account of the state-of-the-art of IFN-α administration as a single agent in melanoma patients. In past experiences, the weak clinical response of melanoma patients to IFNs has hampered the evaluation of the possible mechanisms of their responsiveness. More recently, some clinical trials have been conducted or initiated in an attempt to evaluate hypotheses on the mechanism of IFN-α activity. It may be useful to recall that trials using IFN-α2a or IFN-α2b as single agents in metastatic melanoma yielded a response rate (RR) of approximately 20% with some long lasting responses (Table 2). Up to now, such a limited RR could not be substantially increased by any other combination of IFN-α with cytokines (including IL-2 or TNF-α) in patients with advanced disease [49]. Even combinations of IFN-α2a and DTIC, which showed promising results in early trials [50-52], or IFN-α2a and cis-platinum, need more convincing confirmation by the ongoing randomised trials [53].

More interesting are some recently reported results on the sequential administration of polychemotherapy (BCNU, cis-platinum, DTIC) and biotherapy with IL-2 and IFN-α2 which resulted in a RR of 50% in advanced patients, although with significant toxicity [54].

Conclusions

The biological modulation of melanoma by IFN remains a field of active investigation. The new discoveries regarding the molecular pathways by which IFNs exert their pleiotropic effects on normal and neoplastic cells help to understand and to reinterpret previous laboratory and clinical findings. The number of genes whose activity is modulated by IFNs remains conspicuous, still making it difficult to analyse the effect of such cytokines on each biological function of tumour cells. While it has been confirmed that melanoma cells are highly affected by IFNs in terms of cell surface molecule expression, which may result in a change in the susceptibility to the immune cells or other homeostatic mechanisms, an entirely new aspect of IFN's action emerges, namely the effect of these cytokines on the autocrine or paracrine loops that melanoma cells tend to develop during growth and invasion [28,30]. Once these loops are better defined and their potential role in determining the growth and progression established, new clinical protocols could be designed with an entirely new rationale. On the other hand, a more detailed exploration of the long-term effect of even low doses of IFN on the immune system of patients is also necessary, in particular to ultimately establish the function, if any, of the T cells in controlling melanoma growth *in vivo*. The apparent effectiveness of sequential poly-chemotherapy and biotherapy raises new hopes but also new questions regarding the complex mechanism involved in such a high response rate.

REFERENCES

1   Balch CM, Houghton AN, Milton GW, Sober AJ, Soong SJ: Cutaneous Melanoma. 2nd ed, JB Lippincott Co, Philadelphia 1992

2   Parmiani G, Anichini A, Fossati G: Cellular immune response against autologous human malignant melanoma: are in vitro studies providing a framework for a more effective immunotherapy? JNCI 1990 (82):361-370

3   Garbe C, Krasagakis K, Zouboulis CC, Schröder K, Krüger S, Stadler R, Orfanos CE: Antitumor activity of interferon α, β and γ and their combinations on human melanoma cells in vitro: changes of proliferation, melanin synthesis, and immunophenotype. J Invest Dermatol 1990 (95s):231-237

4   Johns TG, Mackay IR, Callister KA, Hertzog PJ, Devenish RJ, Linnane AW: Antiproliferative potencies of interferons on melanoma cell lines and xenografts: higher efficacy of interferon β. JNCI 1992 (84):1185-1190

5   Trotta PP, Harrison SD: Evaluation of the antitumor activity of recombinant human γ-interferon employing human melanoma xenografts in athymic nude mice. Cancer Res 1987 (47):5347-5353

6   Gomi K, Akinaga S, Oka T, Morimoto M: Analysis of receptors, cell surface antigens, and protein in human melanoma cell lines resistant to human recombinant β- or γ- interferon. Cancer Res 1986 (46):6211-6216

7   Mortarini R, Belli F, Parmiani G, Anichini A: Cytokine-mediated modulation of HLA-class II, ICAM-1, LFA-3 and tumor-associated antigen profile of melanoma cells. Comparison with anti-proliferative activity by rIL1-β, rTNF-α, rIFN-γ, rIL4 and their combinations. Int J Cancer 1990 (45):334-341

8   Ruggiero V, Tavernier J, Fries W, Baglioni C: Induction of the synthesis of tumor necrosis factor receptors by interferon-γ. J Immunol 1986 (36):2445-2450

9   Wallich R, Bulbuc N, Hämmerling GJ, Katzav S, Segal S, Fontana M: Abrogation of metastatic properties of tumour cells by de novo expression of H-2K antigens following H-2 gene transfection. Nature 1985 (315):301-305

10 Van Duinen SG, Ruiter DJ, Bröcker EB, Sorg G, Welvaart K, Ferrone S: Association of low level of class I or high level of class II major histocompatibility complex antigens in metastatic melanoma with high grade of malignancy. J Invest Dermatol 1984 (82) 558-563

11 Bröcker EB, Suter L, Brüggen J, Ruiter DJ, Macher E, Sorg C: Phenotypic dynamics of tumor progression in human malignant melanoma. Int J Cancer 1985 (36):29-35

12 Taramelli D, Fossati G, Mazzocchi A, Delia D, Ferrone S, Parmiani G: Classes I and II HLA and melanoma-associated antigen expression and modulation on melanoma cells isolated from primary and metastatic lesions. Cancer Res 1986 (46):433-439

13 Parmiani G, Anichini A, Castelli C, Fossati G: HLA class II antigens on melanoma cells. In: Ferrone S (ed) Human Melanoma from Basic Research to Clinical Application. Springer-Verlag, New York 1990 pp 197-211

14 Maio M, Gulwani B, Langer JA, Kerbel RS, Duigou GJ, Fisher PB, Ferrone S: Modulation by interferons of HLA antigen, high-molecular-weight melanoma-associated antigen, and intercellular adhesion molecule 1 expression by cultured melanoma cells with different metastatic potential. Cancer Res 1989 (49):2980-2987

15 Nisticó P, Tecce R, Giacomini P, Cavallari A, D'Agnano I, Fisher PB, Natali PG: Effect of recombinant human leukocyte, fibroblast, and immune interferons on expression of class I and II major histocompatibility complex and invariant chain in early passage human melanoma cells. Cancer Res 1990 (50):7422-7429

16 Anichini A, Castelli C, Sozzi G, Fossati G, Parmiani G: Differential susceptibility to recombinant interferon-γ-induced HLA-DQ antigen modulation among clones from a human metastatic melanoma. J Immunol 1988 (140):183-191

17 Vanky F, Wang P, Patorroyo M, Klein E: Expression of the adhesion molecule ICAM-1 and major histocompatibility class I antigens on human tumor cells is required for their interaction with autologous lymphocytes. Cancer Immunol Immunother 1990 (31):19-27

18 Anichini A, Mortarini R, Alberti S, Mantovani A, Parmiani G: T cell receptor engagement and tumor ICAM-1 upregulation are required to bypass low susceptibility of melanoma cells to autologous CTL-mediated lysis. Int J Cancer 1993 (53):994-1001

19 Anichini A, Mortarini R, Supino R, Parmiani G: Human melanoma cells with high susceptibility to cell-mediated lysis can be identified on the basis of ICAM-1 phenotype, VLA profile and invasive ability. Int J Cancer 1990 (46):508-515

20 Johnson JP, Stade BG, Holzmann B, Schwable W, Riethmuller G: De novo expression of intercellular adhesion molecule 1 in melanoma correlates with increased risk of metastasis. Proc Natl Acad Sci 1989 (86):641-644

21 Natali P, Nicotra MR, Cavaliere R, Bigotti A, Romano G, Temponi M, Ferrone S: Differential expression of intercellular adhesion molecule I in primary and metastatic melanoma lesions. Cancer Res 1990 (50):1271-1278

22 Mortarini R, Anichini A, Parmiani G: Heterogeneity for integrin expression and cytokine-mediated VLA modulation can influence the adhesion of human melanoma cells to extracellular matrix proteins. Int J Cancer 1991 (47):551-559

23 Pandolfi F, Trentin L, Lenora AB, Stamenkovic I, Byers HR, Colvin RB, Kurnick JT: Expression of cell adhesion molecules in human melanoma cell lines and their role in cytotoxicity mediated by tumor-infiltrating lymphocytes. Cancer 1992 (69):1165-1173

24 Murray JE, Stuckey S, Pillow JK, Rosenblum MG, Gutterman JU: Differential in vitro effects of recombinant α interferon and recombinant γ interferon alone or in combination on the expression of melanoma associated surface antigens. J Biol Response Mod 1988 (7):152-161

25 Vlock DR, Toporowicz A, Arnold B: Modulation by interferon α and γ of the expression of a melanoma-associated antigen detected by autologous antibody. Melanoma Res 1992 (2):105-114

26 Rosenblum MG, Lamki LM, Murray JL, Carlo DJ, Gutterman JU: Interferon-induced changes in pharmacokinetics and tumor uptake of 111In-labeled antimelanoma antibody 96.5 in melanoma patients. JNCI 1988 (80):160-165

27 Kameyama K, Tanaka S, Ishida Y, Hearing VJ: Interferons modulate the expression of hormone receptors on the surface of murine melanoma cells. J Clin Invest 1989 (83):213-221

28 Herlyn M: Human melanoma: development and progression. Cancer Met Rev 1990 (9):101-112

29 Colombo MP, Maccalli C, Mattei S, Melani C, Radrizzani M, Parmiani G: Expression of cytokine genes, including IL-6, in human malignant melanoma cell lines. Melanoma Res 1992 (2):181-189

30 Kerbel RS: Expression of multi-cytokine resistance and multi-growth factor independence in advanced stage metastatic cancer. Malignant melanoma as paradigm. Am J Pathol 1992 (141):519-524

31 Vedantham S, Gamliel H, Golomb HM: Mechanisms of interferon action in hairy cell leukemia: a model of effective cancer biotherapy. Cancer Res 1992 (52):1056-1066

32 Gruss HJ, Brach MA, Mertelsmann RH, Herrmann F: Interferon-γ interrupts autocrine growth mediated by endogenous IL6 in renal cell carcinoma. Int J Cancer 1991 (49):770-773

33 Maio M, ·Gulwani B, Tombesi S, Ferrone S: Modulation by cytokines of HLA antigens, intercellular adhesion molecule 1 and high molecular weight melanoma associated antigen expression and of immune lysis of clones derived from the melanoma cell line MeM 50-10. Cancer Immunol Immunother 1989 (30):34-42

34 Giacomini P, Aguzzi A, Pestka S, Fisher PB, Ferrone S: Modulation by recombinant DNA leukocyte (a) and fibroblast (b) interferons of the expression and shedding of HLA- and tumor-associated antigens by human melanoma cells. J Immunol 1984 (133):1649-1655

35 Becker JC, Dummer R, Hartmann AA, Burg G, Schmidt R: Shedding of ICAM-1 from human melanoma cell lines induced by IFN-γ and tumor necrosis factor-α. J Immunol 1991 (147):4398-4401

36 Harning R, Mainolfi E, Bystryn J-C, Henn M, Merluzzi VJ, Rothlein R: Serum levels of circulating intercellular adhesion molecule 1 in human malignant melanoma. Cancer Res 1991 (51):5003-5005

37 Huddlestone JR, Merigan TC, Oldstone MB: Induction and kinetics of natural killer cells in humans following interferon therapy. Nature 1979 (282):417-419

38 Brunda MJ, Rosenbaum D, Stern L: Inhibition of experimentally-induced murine metastase by recombinant alpha interferon: correlation between the modulatory effect of interferon treatment on natural killer cell activity and inhibition of metastases. Int J Cancer 1984 (34):421-426

39 Varesio L, Blasi E, Thurman GB, Talmadge JE, Wiltrout RH, Herberman RB: Potent activation of mouse macrophages by recombinant interferon γ. Cancer Res 1984 (44):4465-4469

40 Landolfo S, Cofano F, Giovarelli M, Peat M, Cavallo G, Forni G: Inhibition of IFN-γ may suppress allograft reactivity by T-lymphocytes in vitro and in vivo. Science 1985 (229):176-179

41 Nishimura J, Mitsui K, Ishikawa T, Tanaka Y, Yamamoto R, Suhara Y, Ishizuka H: Antitumor and antimetastatic activities of human recombinant interferon α A/D. Clin Expl Met 1985 (3):295-304

42 Sakurai M, Jigo M, Tamura T, Otsu A, Sasaki Y, Nakano H, Nakagawa K, Minato K, Ohe Y, Saijo N: Comparative study of the antitumor effect of two types of murine recombinant interferons, (β) and (γ), against B16-F10 melanoma. Cancer Immunol Immunother 1988 (26):109-113

43 Mc Millan TJ, Rao J, Everett CA, Hart IR: Interferon-induced alterations in metastatic capacity, class-I antigen expression and natural killer cell sensitivity of melanoma cells. Int J Cancer 1987 (40):659-663

44 Zöller M, Strubel A, Hämmerling G, Andrighetto G, Raz A, Ben-Ze'ev A: Interferon-γ treatment of B16 melanoma cells: opposing effects for non-adaptive and adaptive immune defense and its reflection by metastatic spread. Int J Cancer 1988 (41):256-266

45 Storkus WJ, Howell DN, Salter RD, Dawson J, Cresswell P: NK susceptibility varies inversely with target cell class I HLA antigen expression. J Immunol 1987 (138):1657-1659

46 Schantz SP, Brown BW, Lira E, Taylor DL, Beddingfield N: Evidence for the role of natural immunity in the control of metastatic spread of head and neck cancer. Cancer Immunol Immunother 1987 (25):141-145

47 Hersey P, Edwards A, Milton GW, McCarthy WH: No evidence for an association between natural killer cells activity and prognosis in melanoma patients. Nat Immunol Cell Growth Regul 1983/84 (3):87-94

48 Gansbacher B, Zier K, Cronin K, Hantzopoulos PA, Bouchard B, Houghton AN, Gilboa E, Golde D: Retroviral gener transfer induced constitutive expression of interleukin-2 or interferon-γ in irradiated human melanoma cells. Blood 1992 (80): 2817-2825

49 Kirkwood JM: Studies of interferons in the therapy of melanoma. Sem Oncol 1991 (185):83-90

50 McLeod GRC, Thomson DB, Hersey P: Recombinant interferon α-2a in advanced malignant melanoma. A phase I-II study in combination with DTIC. Int J Cancer 1987 (15):31-35

51 Bajetta E, Negretti E, Giannotti B, Brogelli L, Brunetti I, Sertoli MR, Bernengo MG, Sofra MC, Maifredi G, Zurmiani G, Cornella G, Buzzoni R, Di Leo A, Criscuolo D, Massimiani G, Cascinelli N: Phase II study of interferon α-2a and dacarbazine in advanced melanoma. Am J Clin Oncol 1990 (13): 405-409

52 Hersey P, McLeod GRC, Thomson DB: Treatment of advanced malignant melanoma with recombinant interferon α-2a in combination with DTIC: long-term follow-up of two phase studies. Br J Haematol 1991 (79s):60-66

53 Garbe C, Kreuse H-D, Zouboulis CC, Stadler R, Orfanos CE: Combined treatment of metastatic melanoma with interferons and cytotoxic drugs. Sem

Oncol 1992 (19, S4):63-69
54 Richards JM, Mehta N, Ramming K, Skosey P: Sequential chemoimmunotherapy in the treatment of metastatic melanoma. J Clin Oncol 1992 (10):1338-1343

Head and Neck Cancer

H. Bier [1] and W. Bergler [2]

1 Department of Otorhinolaryngology, Head and Neck Surgery, Heinrich-Heine-University, Moorenstraße 5, 40225 Düsseldorf 1, Germany
2 Klinikum Mannheim, Theodor-Kutzer-Ufer, 68167 Mannheim, Germany

Head and neck cancer accounts for approximately 7% of all incident malignant neoplasms worldwide, and tumours of the mouth and pharynx alone represent the fourth most frequent malignancy in males [1].

The effectiveness of surgery and/or radiation therapy in curing patients with head and neck cancer is well established. These therapeutic measures have resulted in satisfactory survival rates for locally confined tumours with absent or only minor lymph-node metastasis. However, despite radical surgery combined with major plastic functional reconstruction and the development of refined radiotherapeutic techniques, the prognosis of advanced head and neck cancer remains poor. After initial treatment of resectable stage III and IV disease, 50% to 60% of patients have locoregional reccurence and 15% to 25% develop distant metastases.

Therefore, additional treatment modalities are required for the eradication of minimal residual disease [2]:

1. Chemotherapy of squamous cell head and neck carcinomas often leads to dramatic tumour regressions and, particularly in the neoadjuvant setting, complete response rates of 20% to 50% were reported for effective protocols. In general, however, the duration of response is limited to a few months and further chemotherapy is inefficient due to the phenomenon of drug resistance. As yet, chemotherapy has not succeeded in improving the survival of head and neck cancer patients.

2. Cancer biology has revealed new insights into mechanisms of altered differentiation and growth control in neoplastic cells. Use or modulation of proteins and peptides involved in these regulatory networks may open new perspectives for treatment of manifest disease and prevention of disease.

3. Since the immune system is characterised by the high specificity of its recognition and defence mechanisms, immunotherapy represents an attractive approach for the treatment of small tumour burdens that escape conventional treatment.

Immunology of Head and Neck Cancer

Immunological deficits are a common finding in head and neck cancer patients, and they have been demonstrated in numerous *in vitro* and *in vivo* investigations [3,4]. However, the establishment of relevant immunological parameters has been severely limited by the considerable variety of results in this heterogeneous population [5,6]. Incidence and extent of impaired immune reactivity appear to be associated with tumour burden, treatment modality, aetiological factors like smoking and alcohol intake, nutritional status, and age. Interestingly, studies in head and neck cancer patients following successful tumour eradication frequently demonstrate persistent impairments in parameters of immune reactivity similar to untreated patients.

In general, lymphocytes are the most prominent cellular component of inflammatory infiltrates in head and neck carcinomas, but as yet the significance of this lymphocytic reaction remains controversial. More recent investigations have analysed both phenotype and function of these tumour-infiltrating lymphocytes [7-9].

T-lymphocytes make up the major component, and compared to autologous peripheral blood lymphocytes there is a significant increase in activated T cells as judged by the expression of HLA-DR antigens and interleukin-2 receptors [10]. Unfortunately, freshly isolated tumour infiltrating lymphocytes were found to be poor *in vitro* effectors of antitumour cytotoxicity, and they exhibited profound unresponsiveness to different activating agents. However, Cozzolino et al. [11] have been able to isolate T lymphocytes from tumour-invaded lymph nodes of patients with laryngeal cancer that could be induced to proliferate specifically in response to autologous tumour cells. This *in vitro* reactivity required supplementation of interleukin 2 and removal of a distinct supressor T cell population. Very recently, the group of Theresa Whiteside succeeded in generating specific cytotoxic T lymphocytes against an autologous head and neck squamous carcinoma cell line [12].

Further research, in particular functional studies, is needed for a better understanding of both the nature and the aetiology of the multifactorial tumour-host relationship in head and neck cancer patients.

Preclinical Studies with Interferons in Head and Neck Cancer

Interferons were the first cytokines to be used in human cancer. Although effective in a restricted range of malignancies mainly of haematopoietic origin, they appear to be less effective in most solid tumours.

Three major modes of interaction between interferons and neoplastic cells will be discussed here:

1. Interferons modulate growth, differentiation, antigenicity, and response to other regulating factors of tumour cells.
2. Interferons are able to enhance the immune response of the host to the tumour.
3. Apart from their growth regulating and immunomodulating effects, interferons condition tumour cells for the antineoplastic effects of other treatment modalities, e.g. chemotherapy.

For squamous cell head and neck cancer the pleiotropic action of interferons have been in-

Table 1. Mechanisms of action of interferons in head and neck cancer cells

- Growth inhibition
- Differentiation
- Cell cycle alteration
- MHC antigen expression
- Receptor alteration
- Immunomodulation
- Drug modulation

vestigated in some preclinical studies (Table 1).

Growth Inhibition

Interferons appear to enhance the transcription of a large number of genes, at least a few of whose products may be responsible for mediating the antiproliferative effects.

Both antiproliferative and cytotoxic activity of interferons have been described for various squamous carcinoma cell cultures. In the human laryngeal squamous carcinoma cell line HLac 79, (non-recombinant) β-interferon was found to exert the best antiproliferative activity, followed by IFN-α and IFN-γ (Fig. 1). The required concentration of IFN-β to yield an antiproliferative effect of about 50% is clinically achievable by intravenous application. According to Figure 1, a concentration of more than 1,000 IU/ml IFN-α should be necessary for relevant inhibition of proliferation but this serum concentration can hardly be realised or maintained for a sufficient period of time.

In order to observe cytotoxic activity, long-term exposure to interferon exceeding 48 hours is required, and we found a dose-dependent onset of cytotoxicity with IFN-β (Fig. 2). Concentrations over 50 IU/ml IFN-β in the growth medium of cultured HLac 79 cells showed cytotoxicity only after an initial phase of mere antiproliferative action. Lower concentrations did not show the shift of growth inhibition to cytotoxic effect even after long-term incubation.

Short incubation intervals up to 48 hours revealed subsequent recovery of cell proliferation with all interferons tested (Fig. 3). Only IFN-β led to an irreversible inhibition of proliferation at a concentration of more than 1,000 IU/ml. The same effect with IFN-γ or IFN-α would require clinically unfeasible dosages.

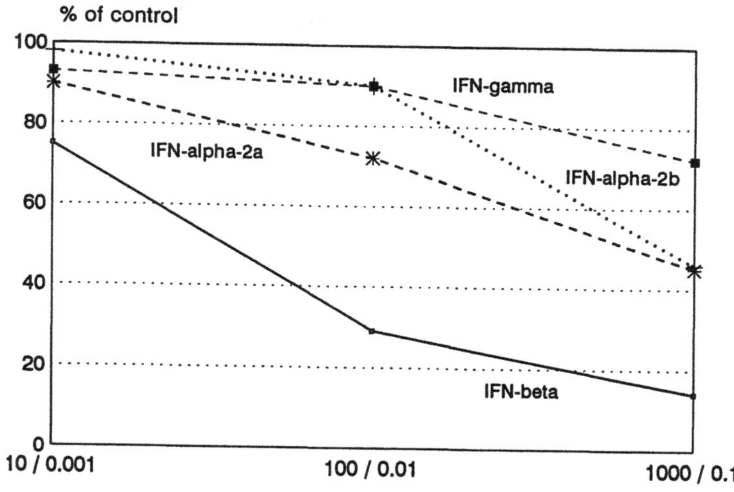

Fig. 1. Dose-response curves for the antiproliferative effect of interferons: proliferation of HLac 79 squamous carcinoma cells measured by BrdU-uptake after 48 hours incubation with interferon-γ (0.001 to 0.1 mg/ml), interferon-β (10 to 1,000 IU/ml), interferon-α2a/2b (10 to 1,000 IU/ml), and untreated control

Richtsmeier et al. [13] assessed the antineoplastic effect of IFN-γ by the loss of cell protein mass by staining cultures with naphtol blue-black dye. The epidermoid carcinoma cell lines JHU-011, SCC-L-P, UM-SCC-16, and A431 were tested, and long-term treatment up to 4 days appeared to be an important factor for the generation of cytotoxic effects.

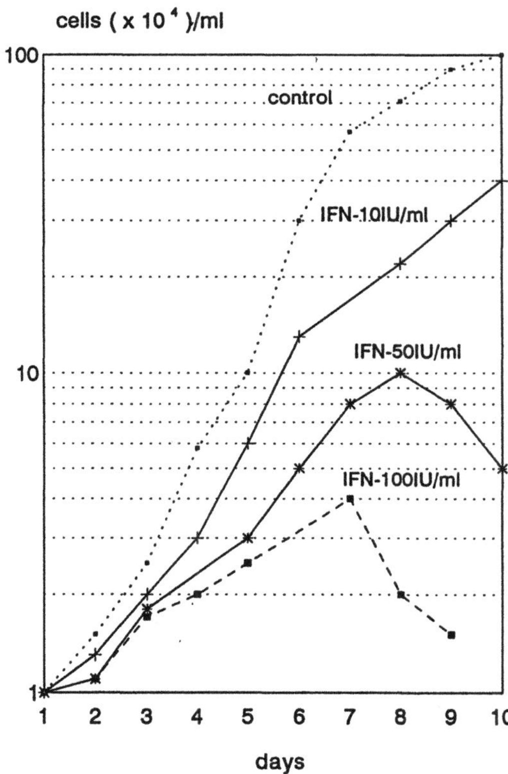

Fig. 2. Growth of HLac 79 monolayer cultures under permanent (non-recombinant) interferon-β (10, 50, and 100 IU/ml) exposure over a period of 10 days with an initial seeding density of 10,000 cells/ml on day 0

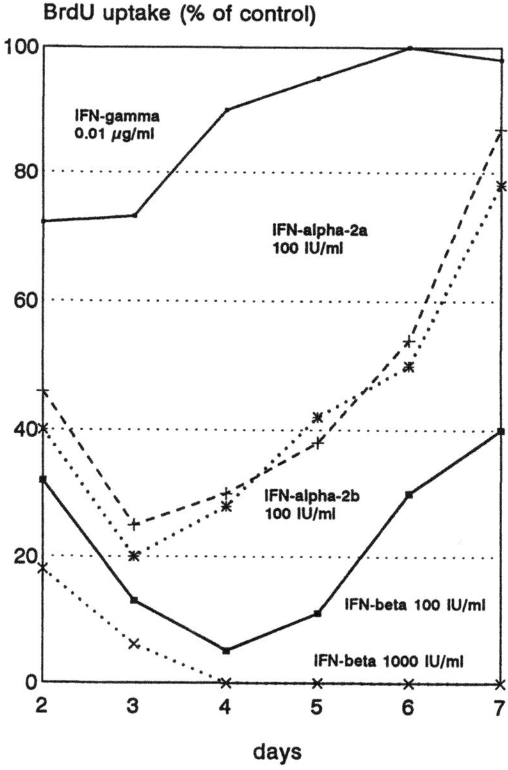

Fig. 3. Effect of 48 hours' interferon treatment (IFN-α2a, 2b 100 IU/ml, IFN-β 100 and 1,000 IU/ml, and IFN-γ 0.01 mg/ml) on HLac 79 cells, and subsequent *in vitro* culture in interferon-free growth medium. In order to monitor cell recovery, proliferation was measured daily by BrdU-uptake for 7 days

In the squamous carcinoma cell line MDA 886 Ln, IFN-β and IFN-γ exhibited a dose-dependent growth inhibition with IFN-γ being the more effective cytokine [14]. However, the same tumour cells grown as multicellular spheroids showed considerable resistance to both interferons.

Interferons were also tested in an epithelial nasopharynx carcinoma hybrid cell line, revealing that IFN-β (10,000 IU/ml) showed better antiproliferative activity than IFN-α (10,000 IU/ml) [15].

Zenner and Zimmermann [16] tried to improve the cytotoxic effect of IFN-γ by simultaneous application of tumour necrosis factor alpha (TNF-α). They treated 5 head and neck carcinoma cell lines with high-dose IFN-γ (0.1 mg/ml) and TNF-α (0.052 mg/ml), which resulted in synergistic action of both cytokines.

Using a colorimetric tetrazolium salt-based MTT assay, Sacchi and coworkers determined the antiproliferative effect of various cytokines on 12 head and neck squamous carcinoma cell lines [17]. All cell lines turned out to be sensitive to IFN-γ with a 50% inhibitory dose ranging between 0.07 and 104 IU/ml. Only one cell line turned out to respond to IFN-α. Again, the combination of interferon with TNF-α showed synergistic antiproliferative activity.

Another approach to increase the antiproliferative action of interferon in a laryngeal squamous carcinoma cell line was performed by Park et al. [18]. IFN-α (1,000 IU/ml) and IFN-γ (100 IU/ml) were tested at 37°C and 39°C for their growth inhibition. Hyperthermia markedly increased the antiproliferative effect of IFN-γ, but no intensified effect was seen with IFN-α.

MHC Antigen Expression

Major histocompatibility (MHC) antigens are essential for immunological recognition, as T-helper cells recognise processed antigen plus MHC class II molecule on antigen-presenting cells, and (most) cytotoxic T cells recognise antigen plus MHC class I molecule on target cells.

The majority of head and neck squamous carcinoma cell lines do not express detectable levels of HLA-DR antigen, but can be induced to display the antigen when treated with low levels of IFN-γ [19,20].

In the squamous carcinoma cell line MDA 886 Ln, both IFN-β and IFN-γ led to MCH class I antigen expression, while only IFN-γ was able to induce MHC class II antigen [14].

These results are in accordance with findings in other *in vitro* tumour models.

Immunomodulation

The induction of class I and class II MHC antigens through interferons, thereby facilitating the recognition of tumour by T cells, has already been mentioned.

In addition, the immunomodulatory effects of interferons can induce or enhance the cytotoxic activity of macrophages, natural killer cells, neutrophils, and T-lymphocytes. For example, lymphokine-activated killer cell generation from peripheral blood lymphocytes of head and neck cancer patients has been significantly improved by the addition of IFN-α [21]. Lymphocytes were incubated with 1,000 IU/ml interleukin 2, 100 IU/ml interleukin 2, 100 IU/ml IFN-α, or 100 IU/ml of both cytokines over a period of 4-5 days. Combined *in vitro* exposure to low-dose interleukin 2 plus IFN-α led to successful lymphokine-activated killer cell generation in a higher percentage of patients. Using a standard 4-hour chromium-51 release assay with Raji tumour target cells, lymphokine-activated killer cell activity was shown to be equivalent to high-dose interleukin 2 stimulation.

Rabinowich et al. [22] analysed the mechanisms that are responsible for interleukin 2-activated natural killer cell-mediated tumour regression of head and neck carcinoma xenografts in thymus-aplastic nude mice. Locally injected NK cells became activated *in situ*, and they produced various cytokines subsequently. In particular IFN-γ appeared to contribute to the observed growth inhibitory effect as determined by the application of neutralising antibodies. Finally, peritumoural injection of IFN-γ alone caused significant growth inhibition of squamous cell head and neck carcinoma xenografts.

Differentiation

Cellular proliferation and differentiation appear to be regulated by discrete gene products including hormone-like effector proteins such as interferons [23].

In order to determine the influence of IFN-β and IFN-γ on the differentiation of a human squamous carcinoma cell line, Sacks et al. [14] measured the activity of transglutaminase by quantifying incorporation of radioactively labelled putrescine into casein. Both interferons induced increased transglutaminase activity in monolayer cells, and IFN-β was twice as active as IFN-γ. In contrast, multicellular tumour spheroids were resistant to the differentiating effect of interferons.

Receptor Alteration

Various cell surface receptors, e.g. those for transferrin, insulin, glucocorticoids, and epidermal growth factor, underly the influence of interferons, and this has to be considered as part of the complex growth regulatory network in terms of feedback control [23].

Inferferon-α has been identified to upregulate the epidermal growth factor receptor in the human epidermoid cancer cell lines A431 and KB [24]. However, flow cytometric studies with HLac 79 carcinoma cells have shown a 100% increase in cell surface bound epidermal growth factor receptor after 48 hours' incubation with 100 IU/ml IFN-β, but neither IFN-α nor IFN-γ were able to produce similar receptor alterations (unpublished own observation).

As high EGF-receptor expression is a characteristic feature of squamous cell head and neck carcinomas, its modulation may be of therapeutic value [25].

Cell to cell adhesion is essential for many events occurring in the course of growth control. Intercellular adhesion molecule 1 (ICAM-1) is a cell surface antigen that mediates effector cell adhesion, activation, and function in inflammatory and immunological reactions. Scher et al. [26] showed markedly enhanced ICAM-1 expression in head and neck squamous carcinoma cell lines after treatment with IFN-γ. Induction of antigen expression was already detectable at 10 IU/ml IFN-γ and reached a maximum at 100 IU/ml.

Cell Cycle Alteration

The passage of cells through all phases of the cell cycle can be slowed down by interferons, although in some cells there are more pronounced effects on a particular phase. Recent findings suggest that arrest of the cell cycle is due to the inhibition of proto-oncogene transcription.

The impact of interferons on cell cycle distribution in the human laryngeal squamous carcinoma cell line HLac 79 is shown in Table 2. Interferon-β was most effective in increasing the percentage of S-phase cells.

Drug Modulation

Interferons are able to modulate the activity of several antineoplastic agents, and there are many examples of the potentation of chemotherapy by interferons *in vitro*. In particular, the interaction between IFN-α and 5-fluorouracil has appeard to be of potential clinical interest. Recent studies suggest that both thymidylate synthetase-dependent and independent mechanisms contribute to the synergistic action of combined IFN-α and 5-fluorouracil [27,28].

Kataoka et al. [29] have determinded the enhancement of the antiproliferative activity of antineoplastic agents by interferons in the head and neck carcinoma cell line HEp 2. In a

Table 2. Distribution of cell cycle phases after 24 hours' treatment of HLac 79 squamous carcinoma cells with IFN-β (100 IU/ml), IFN-α2a/2b (100 IU/ml), IFN-γ (0.01 mg/ml), and untreated control. S-phase labelling with BrdU and FITC conjugated BrdU-antibody for measurement with FACS-scan

Phase	Control	IFN-β	IFN-α2a	IFN-α2b	IFN-γ
G0/1	41%	18%	24%	25%	45%
S	40%	79%	68%	69%	44%
G2/M	19%	4%	10%	7%	11%

5-day proliferation assay single-agent IFN-β revealed superior inhibitory activity compared to IFN-α, and combined treatment with interferon and vinblastine or adriamycin showed the best synergistic effects. However, the underlying mechanisms of these *in vitro* observations remain to be elucidated.

Clinical Investigations with Interferons in Head and Neck Cancer

The antineoplastic activity of IFN-α has already been proven in some malignancies, and current knowledge favours low-dose long-term continuous therapy preferentially performed in patients with small tumour burden. Only a few trials using IFN-β have been conducted, and no therapeutic advantage in comparison to IFN-α has been revealed so far. Though IFN-γ undoubtedly has antitumour activity, its superior immunomodulatory capabilities are not reflected in improved efficacy when compared with IFN-α.

To date, treatment experience with interferons in squamous cell head and neck cancer is limited. Interferon-α has been used by intra- or perilesional and intramuscular injection. Ikic et al. [30] produced measurable antineoplastic activity in head and neck epidermoid cancer with locally administered leukocyte interferon. Recently, the same group reported encouraging response rates in basal and squamous cell carcinomas of the skin [31]. Intra/peritumoural injection of human natural leukocytic interferon led to complete tumour remissions in 61 out of 86 patients with basal cell carcinoma, and in 29 out of 45 patients with squamous cell cancer. Recombinant IFN-α cured 14 out of 20, and 4 out of 10 patients, respectively. The authors attribute the therapeutic effect of locally administered interferon to enhanced macrophage activation at the site of tumour growth. Intramuscular IFN-α treatment has appeared to exert moderate antineoplastic activity in squamous cell head and neck cancer patients as well. In 12 patients suffering from advanced disease, Medenica et al. [32] were able to achieve 3 complete and 4 partial responses. However, other investigators using IFN-α did not observe similar or any antitumour activity at all [33]. In a trial with non-recombinant IFN-α, Vlock et al. [34] produced one complete response in 14 patients with recurrent disease, and this particular patient exhibited markedly increased natural killer cell activity after initial treatment.

Recent studies have investigated the effect of IFN-α on the actvity of established chemotherapy regimens in head and neck cancer. When combined with cisplatin and continuous infusion 5-fluorouracil, IFN-α produces response rates comparable to those achieved with chemotherapy alone, but increased toxicity may compromise the treatment schedule [35,36]. Similarly, a Norwegian trial comparing preoperative radiotherapy alone versus combined radiation and IFN-α treatment had to be discontinued due to considerable side effects [37]. As yet, IFN-γ has not been able to produce significant anticancer activity in squamous cell head and neck neoplasms [38].

Conclusion

Since their first description as cytokines that are produced in response to virus infections, interferons are now considered as a protein family exerting pleiotropic regulatory effects on both immune response and control of growth and differentation in normal and neoplastic cells. Very recent findings suggest a significant role for transcription factors of interferon type I in the expression of normal and malignant phenotypes [39,40].

With regard to head and neck cancer, there is evidence that interferons can produce antineoplastic activity *in vitro* and *in vivo*. However, the underlying mechanisms of the observed cytostatic and/or cytotoxic effects largely remain to be established.

REFERENCES

1 Parkin DM, Läärä E, Muir CS: Estimates of the worldwide frequency of sixteen major cancers in 1980. Int J Cancer 1988 (41): 184-197

2 Bier H: Current trends in therapeutic research. Rec Res Cancer Res 1993 (134): 162-170

3 Vlock DR: Immunobiologic aspects of head and neck cancer. Clinical and laboratory correlates. Hematol Oncol Clin North Am 1991 (5): 797-820

4 Wanebo HJ, Jun MY, Strong EW, Oettgen H: T-cell deficiency in patients with squamous cell cancer of the head and neck. Am J Surg 1975 (130): 445-451

5 Bier J, Nicklisch U, Platz H: The doubtful relevance of non-specific immune reactivity in patients with squamous cell carcinoma of the head and neck region. Cancer 1983 (52):1165-1172

6 Parkinson DR, Schantz SP: Immunobiological therapy for head and neck cancer. In: Snow GB and Clark IR (eds) Multimodality therapy for head and neck cancer. Thieme, Stuttgart, New York 1992 pp 147-159

7 Heo DS, Whiteside TL, Johnson IT, Chen K, Barnes EL, Herberman RB: Long term interleukin-2 dependent growth and cytotoxic activity of tumor infiltrating lymphocytes from human squamous cell carcinomas of the head and neck. Cancer Res 1987 (47): 6353-6362

8 Miescher S, Whiteside TL, Carrel S, Fliedner V von: Functional properties of tumor infiltrating and blood lymphocytes in patients with solid tumors: effect of tumor cells and their supernatents on proliferative responses of lymphocytes. J Immunol 1986 (136): 1899-1907

9 Snyderman CH, Heo DS, Chen K: T-cell markers in tumor-infiltrating lymphocytes of head and neck cancer. Head Neck 1989 (11): 331-337

10 Whiteside TL, Heo DS, Takagi S, Herberman RB: Tumor-infiltrating lymphocytes from human solid tumors: antigen-specific killer T lymphocytes of activated natural killer lymphocytes. In: Stevenson HC (ed) Adoptive cellular immunotherapy of cancer. Marcel Dekker, New York 1989 pp 139-157

11 Cozzolino F, Torcia M, Carossino AM, Giordani R, Selli C, Talini G, Ferrarini M: Characterization of cells from invaded lymph nodes in patients with solid tumors. J Exp Med 1987 (166): 303-315

12 Hirabayashi H, Yasumura S, Schwartz D, Johnson J, Whiteside T: Generation of cytotoxic T lymphocytes against autologous squamous cell carcinoma of the head and neck. Proc International Conference Head and Neck Cancer, Elsevier, Amsterdam 1992 (3): 129

13 Richtsmeier WJ: Interferon gamma induced oncolysis: an effect on head and neck squamous carcinoma cultures. Arch Otolaryngol Head Neck Surg 1988 (114): 432-437

14 Sacks PG, Racz T, Schantz SP, Rosenblum MG: Growth inhibition by beta and gamma interferon of MDA 886 Ln monolayer cells and multicellular tumor spheriods: a differentiation therapy model for squamous cell carcinoma. Proc International Conference Head and Neck Cancer, Elsevier, Amsterdam 1992 (3): 111

15 Takimoto T, Ishikawa S, Umeda R, Ogura H: Effects of human interferon on cell proliferation and antigens induced by Epstein-Barr virus in epithelial hybrid cells derived from nasopharyngeal carcinoma. Otoralyngol Head Neck Surg 1985 (93): 500-504

16 Zenner HP, Zimmermann U: Cytotoxic effects of recombinant human biological response modifiers on head and neck carcinoma cells. ORL 1988 (30): 150-155

17 Sacchi M, Klapan I, Jonson JT, Whiteside TL: Antiproliferative effects of cytokines on squamous cell carcinoma. Arch Otolaryngol Head Neck Surg 1991 (117): 321-326

18 Park RI, Richtsmeier WJ: Hyperthermia effects on the growth of a laryngeal squamous cell carcinoma line treated with recombinant human interferons alpha and gamma. Otolaryngol Head Neck Surg 1989 (101): 542-548

19 Houck JR, Shah TP, Ohlsson-Wilhelm BM, Kloszewski E, Conner BR: Modulation of human leukocyte antigen class I expression by gamma interferon in head and neck cancer cell lines. Am J Otolaryngol 1988 (9): 217-223

20 Koch WM, Scher RL, Richtsmeier WJ: Cytokine modulation of HLA-DR expression of squamous cell carcinoma. Proc Int Conference Head and Neck Cancer, Elsevier, Amsterdam 1992 (3): 190

21 Wanebo H, Blackinton D, Weigel T, Turk P, Mehta S: Augmentation of the lymphokine-activated killer cell response in head and neck cancer patients by combination interleukin-2 and interferon alpha. Am J Surg 1991 (162): 384-387

22 Rabinowich H, Vitolo D, Altarac S, Herberman RB, Whiteside TL: Role of cytokines in the adoptive immunotherapy of an experimental model of human head and neck cancer by human IL-2 activated natural killer cells. J Immunol 1992 (149): 340-349

23 Kumar R: Interferons Proc Am Assoc Cancer Res 1992 (33): 581-582

24 Budillon A, Tagliaferri P, Caraglia M, Torrisi MR, Normanno N, Iacobelli S, Palmieri G, Stoppeli MP, Frati L, Bianco AR: Upregulation of epidermal growth factor receptor induced by alpha interferon in human epidermoid cancer cells. Cancer Res 1991 (51): 1294-1299

25 Bergler W, Bier H, Ganzer U: Cisplatin induced EGF receptor reduction in malignant cells. ORL 1990 (52): 297-302

26 Scher RL, Koch WM, Richtsmeier WJ: Expression of the intercellular adhesion molecule (ICAM-1) on squamous cell carcinoma. Proc International Conference Head and Neck Cancer, Elsevier, Amsterdam 1992 (3): 191

27 Seymour MT, Dobson N, Clemens M, Slevin ML: 5-Fluorouracil / interferon-alpha synergy: is regulation of expression of thymidylate synthetase the key? Proc Am Assoc Cancer Res 1992 (33): 545

28 Wadler S, Mao X, Schwartz EL: Recombinant alpha-2a-interferon augments 5-fluorouracil effects on nucleotide pools and DNA double strand breaks in human colon cancer cell lines. Proc Am Assoc Cancer Res 1992 (33): 425

29 Kataoka T, Oh-Hashi F, Sakurai Y: Enhancement of antiproliferative activity of vinblastine and adriamycin by interferon. Gann 1984 (75): 548-556

30 Ikic D, Padovan I, Brodared I, Knezevic M, Soos E: Application of human leucocyte interferon in patients with tumors of the head and neck. Lancet 1981 (1): 1025-1027

31 Ikic D, Padovan I, Pipic N, Knezevic M, Djakovic N, Rode N, Kosutic I, Belicza M, Cajkovac V: Interferon therapy for basal cell carcinoma and squamous cell carcinoma. Int J Clin Pharmacol Ther Toxicol 1991 (29): 342-346

32 Medinca R, Slack N: Clinical results of leukocyte interferon-induced tumor regression in human metastatic cancer resistant to chemotherapy and/or radiotherapy - pulse therapy schedule. Cancer Drug Deliv 1985 (2): 53-75

33 Miyake M, Horuchi M, Togowa K, Nobujara K, Kimoto T: Recombinant interferon alpha 2 in patients with head and neck cancer. Gan to Kagaku Ryoho 1985 (12): 1651-1655

34 Vlock DR, Johnson J, Myers E, Day R, Gooding WE, Whiteside T, Pelch K, Sigler B, Wagner R, Colao D, et al: Preliminary trial of nonrecombinant interferon alpha in recurrent squamous cell carcinoma of the head and neck. Head Neck 1991 (13): 15-21

35 Shirinian M, Choksi AJ, Dimery I, Heyne K, Lippman S, Guillory C, Hong WK: Phase I/II study of cisplatin + 5 fluorouracil + alpha interferon for recurrent squamous cell carcinoma of the head and neck. Proc Am Assoc Clin Oncol 1992 (11): 249

36 Vokes EE, Ratain MJ, Mick R, Wenig B, Weichselbaum RR, Berezin F, Kezloff M, et al.: A clinical and pharmacologic phase I-II study of PFL with interferon alpha-2b for head and neck cancer. Proc Am Assoc Clin Oncol 1992 (11): 243

37 Valavaara R, Kortekangas AE, Nordman E, Cantell K: Interferon combined with irradiation in the treatment of operable head and neck carcinoma. A pilot study. Acta Oncol 1992 (31): 429-431

38 Richtsmeier WJ, Koch WM, McGuire WP, Poole ME, Chang EM: Phase I-II study of advanced head and neck squamous cell carcinoma patients treated with recombinant human interferon gamma. Arch Otolaryngol Head Neck Surg 1990 (116): 1271-1277

39 Willmann CL, Sever CE, Pallavicini MG, Harada M, Tanaka N, Slovak ML, Yamamoto M, et al: Deletion of IRF-1, mapping to chromosome 5 q 31.1, in human leukemia and preleukemic myelodysplasia. Science 1993 (259): 968-971

40 Harada H, Kitagawa M, Tanaka N, Yamamoto M, Harada K, Ishihara M, Taniguchi T: Anti-oncogenic and oncogenic potentials of interferon regulatory factors- 1 and -2. Science 1993 (259): 971-974

Breast Cancer

J.M. Ferrero and M. Namer

Centre Antoine Lacassagne, 36 Voie Romaine, 06054 Nice Cedex, France

An ever-increasing number of biological agents are known to have a significant influence on the proliferation and differentiation of the ductal epithelium of the normal and malignant breast. These biologicals can exert a direct antiproliferative effect on cells, modulate the secretion of growth factors by a tumour, or increase the immune defenses of the host. The strategies employed for the clinical development of these biologicals rely on an in-depth understanding of their effects on tumour biology and immune function.

Biological Approaches to Breast Cancer Therapy

Many different types of hormones influence the growth and differentiation of malignant mammary cells. Amongst these, a new class of agents, the so-called growth factors, have been identified during the last few years: transforming growth factors (TGF-α and β), epidermal growth factor (EGF), and insulin growth factor (IGF). The secretion of these growth factors is partly controlled by cytokines, and in particular by interferons (IFNs) [1].

Although the risk of breast cancer does not appear to be increased by immuno-depression, several observations suggest that immune functions play an important role in the natural history of the disease. Peripheral lymphopenia and lymphocytic infiltration of the primary tumour are associated with a bad prognosis [2]. Spread to lymph nodes and the absence of oestrogen receptor expression are associated with a significant reduction in natural killer (NK) cell activity [3,4]. Activated T-lymphocytes from regional axillary lymph nodes show marked cytotoxicity toward malignant target cells in culture [5]. TGF-β secreted by malignant breast cancer cells potentiates the immune functions of lymphocytes [6]. All these observations suggest that the more immunogenic the tumour and the more effective the immune system of the host, the better the prognosis.

Biological response modifiers can be used to different ends [7]:
- to increase the antiproliferative effects of some hormones and drugs (IFNs);
- to increase cell immunity (IL-2, monoclonal antibodies);
- to increase cell differentiation (retinoids, IFNs);
- to reduce drug toxicity (haematopoietic growth factors).

IFNs are a family of proteins active on immune function and tumour biology. IFNs I and II increase peripheral NK cell activity and macrophage cytotoxicity [1]. The expression of antigens of the major histocompatibility complex (MHC) and of tumour-associated antigens is also increased by IFNs, thus facilitating recognition of tumour cells by the effector immune system of the host or by monoclonal antibodies [8].

Type I IFNs (α and β) have antiproliferative activity, and enhance the cytotoxic action of many drugs on cultures of adenocarcinoma, epidermoid and lymphoid cell lines [9]. IFN-α increases the effectiveness of doxorubicin and cyclophosphamide against a human tumour grafted onto nude mice [10] and potentiates the cytotoxicity of 5-fluorouracil (5-FU) in human mammary tumour cell lines by increasing the stability of the ternary complex which 5-FU forms with thymidylate synthetase [11]. In an *in vivo* tumour model, IFN protected the host from the undesirable toxic effects of 5-FU [10,12].

The first studies on types I and II IFNs concerned cell growth inhibition and steroid hormone receptor regulation in mammary tumour cell lines. After reviewing these data, we shall discuss the clinical results obtained first with monotherapy, then with combination hormone and/or cytotoxic chemotherapy.

Preclinical Studies

IFNs and Receptors

IFNs inhibit the growth of some cell lines *in vitro*. Inhibition of several epidermoid cell lines by IFNs is linked to changes in EGF receptor expression. In a cell line whose growth is dependent upon EGF, IFN-γ decreased the number of EGF receptors [13] without modifying EGF affinity for its receptor, and thereby led to a decrease in cell growth.

Several *in vitro* studies have established a correlation between IFNs and oestrogens. Most teams observe an increase in oestrogen receptor (ER) levels after stimulation of cell cultures with low IFN concentrations [14-16]. In one study [17], a 2 to 7-fold increase in ER was recorded 48 hours after IFN-α addition, while oestrogens, and in particular 17β-oestradiol, decrease IFN-α receptors [18].

The increase in ER concentration by IFNs has naturally led to the study of the action of anti-oestrogens on the growth of IFN-pretreated cells. The anti-oestrogen tamoxifen is a standard treatment for metastatic breast cancer and gives an overall response rate of about 30% which increases to 70% in case of ER-positive tumours. Tamoxifen decreases ER content in MCF7 cells; IFN-α prevents this anti-oestrogen-induced ER reduction. Both tamoxifen and its analogue toremifene increase IFN receptors and exert an additive [17], if not always synergistic, effect [19, 20] with IFN-α on cell growth inhibition. The lack of affinity of either anti-oestrogen for ER is influenced by IFN-α.

The effects of IFNs on progesterone receptor (PR) are more controversial. Since PR is oestrogen induced in most systems, it may not be surprising that a longer time interval was found necessary to increase PR levels by IFN-α [21]. Both progestins and antiprogestins decreased IFN-α receptors in human ZR-75-1 breast cancer cells and variants [18].

IFNs and Tumour Necrosis Factor (TNF)

An increase in cytotoxicity has been observed using combinations of tamoxifen, interferons types I and II, and TNF in MCF7 cells . TNF is an immunomodulator and mediator of monocyte cytotoxicity induced by itself, TFN-γ and interleukin-1 [7].

IFNs and Killer Cells

As mentioned above, types I and II IFNs increase peripheral NK cell activity and macrophage cytotoxicity. NK and LAK (lymphokine-activated killer) cells recognise and destroy their target cells which have reduced expression of class I MHC antigens by a still unknown mechanism. Several studies have suggested that IFNs, and in particular IFN-γ, increase the expression of these antigens, thereby inducing resistance to NK and LAK cells. Resistance was abolished by class I and II anti-human leukocyte antigen (HLA) antibodies. Moreover, IFN-β can reverse the decrease in NK cell activity induced by cytotoxic chemotherapy *in vivo* [22]. This observation may constitute a rational basis for IFN co-administration to avoid the inevitable immune depression of the host caused by chemotherapy. IFN-γ significantly increases the expression of class I and II MHC antigens on several breast cancer cell lines [23]. Comparative *in vitro* studies have shown that only IFN-γ can induce the synthesis of HLA-Dr type membrane antigens [24].

IFNs and Gene Expression

In an ER-positive breast cancer cell line (MCF7 clone), IFNs-α and β induced the messenger RNAs (mRNAs) of several proteins, indicating an increase in the transcription of several genes (651, 6-16, 1-8, 2A), particularly those coding for class I antigens of the HLA system [25]. IFN-γ increased the concentration of the mRNA of the c-myc proto-oncogene product, but inhibited cell growth [26].

Clinical Studies

IFN Monotherapy

The first clinical results with IFN-α from leukocytes in the early 1980s yielded encouraging response rates in patients with metastatic breast cancer, and generated considerable enthusiasm. Response rates of 35% and 22%, but of highly variable duration - from 8 to 60 weeks and 2 to 25 weeks, respectively - were recorded [27,28]. Furthermore, intramuscular injections of fibroblastic IFN were found to significantly decrease skin lesions or induce their marked necrosis and an inflammatory reaction in 10 out of 11 patients with skin metastases from a primary breast tumour [29]. NK cell activity in peripheral blood was considerably increased.

However, these results were not confirmed in most subsequent phase II studies. The combined results of 6 phase II studies with IFN-α yielded response rates of less than 5% (Table 1) and revealed substantial toxicity, frequently requiring considerable dose reduction. Discrepancies between initial results and those in Table 1 could be explained by differences in IFN purity. In the initial studies, IFN was only partially purified and could have contained other active cytokines.

Neither systemic IFN-γ nor IFN-β have demonstrated significant efficacy as monotherapies in patients refractory to all previous hormone or chemotherapies [36,37]. However, IFN-β applied locally to metastatic pleural effusions gave a response rate of 41%, of which 20% were complete responses [38,39].

Table 1. Clinical results with interferon monotherapy in breast cancer

References	IFN-α	Protocols	Responding/Evaluable patients	
Gutterman et al. [28]	Partially purified from leukocytes	3×10^6 IU/m^2 3 times/wk		6/17
Borden et al. [28]		3×10^6 IU/m^2/day		5/23
			Total:	11/40 = 27.5%
Sherwin et al. [30]	IFN-α A recombinant	50×10^6 IU/m^2 3 times/wk		0/17
Silver et al. [31]	Lymphoblast	2×10^6 IU/m^2/day		1/27
Muss et al. [32]	IFN-α 2 recombinant	50×10^6 IU/m^2/day		0/12
Smedley et al. [33]	IFN-α_a recombinant	20×10^6 IU/m^2/day		2/12
Goodwin et al. [34]	Lymphoblast	3×10^6 IU/m^2/day 3 times/wk		1/22
Sarna et al. [35]	Lymphoblastoid	30×10^6 IU/m^2/day		0/15
			Total:	4/105 = 4%

IFNs in Combination with Tamoxifen or Progestins

In vitro tamoxifen decreases ER levels in receptor-positive breast cancer patients, as observed on repeated tissue sampling. *In vitro* IFN can stimulate ER in patients, as shown by the study of skin metastases before and after treatment with IFN-α2b (3 million units/day for 14 days) [40].

IFN has been combined with tamoxifen in several phase II studies. IFN is either administered to patients who have relapsed on tamoxifen therapy or as a first-line therapy in combination with tamoxifen in order to increase the response rate (Table 2). In one of the first open studies [41], 9 patients resistant to tamoxifen received IFN-α (5 MU/m^2, 5 days per week) and tamoxifen (20 mg/day). No partial nor complete responses were observed. Out of the 10 patients who received the same combination as a first-line therapy, 5 responded; of the ER-positive patients in this small group, 71% responded. Thus, there does not seem to be any advantage in administering a combination. In other phase II trials in tamoxifen-resistant patients, the responses varied from 6% to 26% [42-44], the best responses being obtained with IFN-β with a median duration of response of 6 months [42,45].

In one study, a retinoid (retinol palmitate 1500 IU BID), which, like tamoxifen, can induce TGF-β *in vitro*, was administered with IFN-β and tamoxifen to 33 ER$^+$ or ER$^-$ patients who had received no prior tamoxifen therapy. A response rate of 61% was recorded; 22% were complete responses and the mean duration of response was 14 months [46]. A randomised study is ongoing in order to further investigate this combination.

All these results do not suggest that IFNs can reverse tamoxifen resistance nor improve first-line response. Combinations of tamoxifen with other cytokines and retinoids (TFN and retinol palmitate) are currently under investigation.

One study in a small number of patients (n=22) comparing the progestin medroxyprogesterone acetate (MPA) administered alone to MPA plus IFNs-β and γ gave a small, nonsignificant advantage to the combination which, however, was accompanied by a considerable increase in toxicity [47].

IFNs and Cytostatics

Because of the synergy observed *in vitro* between INF-α and some cytostatics, several

Table 2. Clinical results with combined interferon + tamoxifen in breast cancer

References	IFN	No. pts.	Response rates (%)
Miglietta et al. [43]	IFN-α 5 MU x 3/wk	13	15.4
Buzzi et al. [42]	IFN-β 3 MU/d 14 days	33	26
Van Den Berg et al. [17]	IFN-α 3 MU x 3/wk	16	6
Macheledt et al. [41]	IFN-α 5 MU/m2/d 5 d/wk	9* 10**	0 50
Repetto et al. [45]	IFN-β 5 MU/d x 3/wk	26	23

* tamoxifen resistant   ** with no prior treatment

Table 3. Clinical results with combined interferon + chemotherapy in breast cancer

References	IFN	Cytostatic agent	No. pts.	Response rate (%)
Spielman et al. [48]	IFN-α 4.5 MU/d x 3 weeks	5 FU Folinic acid	14	21
Walters et al. [49]	IFN-α 5 MU/m^2 x 3/wk	5 FU	18	11
Pronzato et al. [50]	IFN-α 3 MU/d 14 days	Cisplatin	14	0
Schnalke et al. [51]	IFN-α 5 MU/m^2 x 3/wk	5 U Folinic acid	16	40

phase II studies in patients who had relapsed on prior therapy investigated the possible clinical value of these combinations. The cytostatic drugs used in combination with IFN-α were 5-fluorouracil (5-FU), folinic acid or cis-di-amminodichloroplatinum, which gave highly variable and inconclusive results (Table 3) since the number of patients recruited was very small. IFN-α seemed to enhance the mucous toxicity of 5-FU.

IFNs and Other Cytokines

IFN-α combined with other cytokines, in particular interleukin-2, has shown no great efficacy. Seven patients with metastatic breast cancer received IL-2 (3.10^{-6} IU/m^2/d) in combination with INF-α (5.10^6 IU/m^2/d) x 4 days/week during 4 weeks. All patients were pretreated. One patient showed a minor response on lung metastases, and 2 showed a stabilisation. No partial nor complete response has been observed [52].

IFNs as Adjuvant Therapy

IFN-α has also been used as an adjuvant therapy in breast cancer. Thirty-two patients after treatment of breast cancer recurrence were randomised into 2 groups receiving either no adjuvant therapy or IFN-α (3 million units/day for 1 year). There was no difference in the recorded recurrence rates in the 2 groups [53]. A much larger study initiated in 1982 on 311 patients randomised, after adjuvant therapy with doxorubicin and cyclophosphamide, into 2 groups: no therapy and IFN-α (3 million units/day for 1 year). This study revealed no difference in time to recurrence or survival after 3 years of follow-up [54]. The results seem therefore to indicate that IFN-α has little potential in the adjuvant treatment of breast cancer [55].

In conclusion, the encouraging results obtained with different IFNs *in vitro* on ER modulation and cell growth have not been confirmed in clinical trials. IFNs do not appear to have any significant action, either alone or in combination, on metastatic breast cancer or as adjuvant therapy. Combinations with other cytokines, in particular TNF, and with retinoids may prove to be more promising and some are currently being tested in clinical trials. However, considering the large number of possible combination therapies with different cytokines, cytostatics, and hormones which cannot reasonably be tested in patients, substantial further laboratory research is required to elucidate the mechanisms of action of these agents, allowing to select the most promising combinations.

REFERENCES

1   Goldstein D, Laszlo J: Interferon therapy in cancer: from Imaginon to Interferon. Cancer Res 1986 (46):4315-4329

2   Pattison CW, Woods KL, Morrison JM: Lymphocytopenia as an independent predictor of early recurrence in breast cancer. Br J Cancer 1987 (56):75-76

3   Levy S, Heyberman R, Lippman ME et al: Correlative of stress factors with sustained depression of natural Killer cell activity and prognosis in patients with breast cancer. J. Clin Oncol 1987 (5):348-353

4   Zielinski C, Tichatschek E, Muller C et al: Association of increased lytic effector cell function with high estrogen receptor levels in tumor-bearing patients with breast cancer. Cancer 1989 (63):1985-1989

5   Bonilla F, Alvarez-Mon M, Merino F et al: Interleukin-2 induces cytotoxic activity in lymphocytes from regional axillary nodes of breast cancer patients. Cancer 1988, (61):629-634

6   Espevik T, Waage A, Faxaag LA et al: Regulation of Interleukin-2 and Interleukin-6 production from T-cells: involvement of Interleukin-1 beta and transforming growth factor-beta. Cell Immunol 1990 (126):47-56

7   Sparanano JA, O'Boyle K: The potential role for biological therapy in the treatment of breast cancer. Semin Oncol 1992 (19):333-341

8   Carrel S, Schmidt-Kessen A, Giuffre L: Recombinant Interferon-gamma can induce the expression of HLA-DR and DC- or DR-negative melanoma cells and enhance the expression of HLA-ABC and tumor associated antigens. Eur J Immunol 1985 (15):118-123

9   Wadler S, Schwartz EL: Antineoplastic activity of the combination of Interferon and cytotoxic agents against experimental and human malignancies: a review. Cancer Res 1990 (50):3673-3686

10 Balkwill FR, Moodie EM: Positive interaction between human interferon and Cyclophosphamide or Adriamycin in a human tumor model system. Cancer Res 1984 (44):904-908

11 Wadler S, Schwartz EL, Thompson D et al: Effects of recombinant Interferon or thymidylate synthetase expression and ternary complex formation in human tumor cell lines. Proc Am Soc Clin Oncol 1990 (31):402 (abstract)

12 Stolfi RL, Martin DS, Sawyer RC et al: Modulation of 5-Fluorouracil-induced toxicity in mice with Interferon or with the Interferon inducer, polyinosinic-polycytidylic acid. Cancer Res 1983 (43):561-566

13 Chakravarty A, Chen LC, Mehta D, Hamburger AW: Modulation of epidermal growth factor receptors by gamma Interferon in a breast cancer cell line. Anticancer Res 1991 (11-1):347-351

14 Calvo F, Solary E, Prudhomme JF et al: Gamma Interferon induce an inverse modulation of estrogen receptors in two human breast cancer cell lines: relation with antiproliferative effects and expression of an estrogen regulated protein PS2/BCEI. Proc Am Meet Am Soc Clin Oncol 1988 (7):A128

15 Porzsolt F, Otto AM, Trauschel B, Bick C et al: Rationale for combining Tamoxifen and Interferon in the treatment of advanced breast cancer. J Cancer Res Clin Oncol 1989 (115):465-469

16 Sica G, Iacopino F, Lawa G et al: Natural beta Interferon enhances steroid hormone receptor level in breast and endometrial cancer. J Interferon Res 1989 (9-suppl. 2):S116

17 Van Den Berg HW, Leabey WJ, Lynch M et al: Recombinant human Interferon alpha increases estrogen receptor expression in human breast cancer cells (2R-75-1) and sensitizes them to the anti-proliferative effects of Tamoxifen. Br J Cancer 1987 (55):255-257

18 Martin JHJ, Mc Kibben BM, Lynch M, Van Den Berg HW: Modulation by oestrogen and progestins/antiprogestins of alpha Interferon receptor expression in human breast cancer cells. Eur J Cancer 1991 (27):143-146

19 Kangas L, Nieminen AL, Cantell K: Additive and synergistic effects of a novel antiestrogen, Toremifene, and human Interferons as estrogen responsive MCF-7 cells in vitro. Med Biol 1985 (63):187-190

20 Epstein LB, Benz CC, Doty E: Synergistic antiproliferative effect of Interferon-alpha and Tamoxifen on human breast cancer cells in vitro. Clin Res 1986 (35):196A

21 Sica G, Incopino F, Della Cuna GR, Marchetti P: Recombinant Interferon-alpha 2B effects proliferation, steroid receptors and sensitivity to Tamoxifen of cultured breast cancer cells (CG-5). Anticancer - Drugs 1992 (3):147-153

22 Tentori L, Fugetta MP, D'Atri S et al: Influence of low-dose beta-Interferon on Natural Killer cell activity in breast cancer patients subjected to chemotherapy. Cancer Immunol Immunother 1987 (24):86-91

23 Jabrane-Ferrat N, Faille A, Loiseau P et al: Effect of gamma Interferon on HLA class I and II transcription and protein expression in human breast adenocarcinoma cell lines. Int J Cancer 1990 (45-6): 1169-1176

24 Gastl G, Marth C, Leiter E et al: Effects of human recombinant alpha-2 Arg-Interferon and gamma-Interferon on human breast cancer cell lines: dissociation of antiproliferative activity and induction of HLA-DR antigen expression. Cancer Res 1985 (45):2957-2961

25 Tiwari RK, Rossi JD, Teland NT, Osborne MP: Molecular response of human breast carcinoma cells to Interferon alpha. Proc Ann Meet Am Assoc Cancer Res 1989 (30):A44

26 Hamburger AW, Pinnameni G: Interferon induced increases in c-myc expression in a human breast carcinoma cell line. Anticancer Res 1991 (11):1891-1894

27 Gutterman JV, Blumenschein GR, Alexanian R et al: Leucocyte Interferon-induced tumor regression in human metastatic breast cancer, multiple myeloma and malignant lymphoma. Ann Intern Med 1980 (93):399-401

28 Borden EC, Holland JF, Dao TL et al: Leucocyte derived Interferon (alpha) in human breast

carcinoma. The American Cancer Society Phase II trial. Ann Intern Med 1982 (97): 1-6

29 Pouillart P, Palangie T, Jouve M et al: Administration of fibroblast Interferon to patients with advanced breast cancer: possible effects on skin metastasis and on hormone receptors. J Cancer Clin Oncol 1982 (18):929-935

30 Sherwin SA, Mayer D, Ochs J et al: Recombinant leukocyte A Interferon in advanced breast cancer. Results of a Phase II efficacy trial. Ann Inter Med 1983 (98):598-602

31 Silver HKB, Connors J, Salinas F et al: Treatment response in a prospectively randomized study of high vs low dose treatment with lymphoblastoid Interferon. Proc. ASCO 1983 (2):51

32 Muss H, Caponera M, Stuart J et al: Intravenous Interferon α2 in patients with advanced breast cancer: a phase II study. Proc ASCO 1983 (2):99

33 Smedley H, Katrak M, Sikova K et al: Recombinant human Interferon in advanced breast cancer. Br J Cancer 1983 (47):566-567

34 Goodwin BJ, Brenckman W, Moove J et al: Phase II trial of human lymphoblastoid Interferon (Wellferon) in metastatic breast carcinoma. Proc ASCO 1984 (3):60

35 Sarna GP, Figlin RA: Phase II trial of α-lymphoblastoid Interferon given weekly as treatment of advanced breast cancer. Cancer Treat report 1985 (69):347-349

36 Muss HB, Caponera M, Zekan PJ et al: Recombinant gamma Interferon in advanced breast cancer: a phase II trial. Invest. New Drugs 1986 (4-4):377-381

37 Barreras L, Vogel CL, Koch G, Marcus SG: Phase II trial of recombinant beta-Interferon in the treatment of metastatic breast cancer. Invest New Drugs 1988, 6 (3):211-215

38 Jereb B, Krasovec M, Cervek J, Soos E: Intrapleural application of human leucocyte Interferon (HLI) in breast cancer patients with ipsilateral pleural carcinomatosis. J. Interferon Res 1987 (7):357-363

39 Terzoli E, Izzo F, Lucatelli S, Ranuzzi M: Natural beta-Interferon for local use in the treatment of metastatic pleural effusion caused by neoplasm of the breast. G Ital Oncol 1989 (9):53-57

40 Hakes T, Menendez-Botet C, Moore M, Osborne M: Modulation of estrogen and progesterone receptors in human breast cancer by alpha-2B Interferon. Proc Ann Meet Am Soc Clin Oncol 1990 (9):A167

41 Macheledt JE, Buzdar AV, Hortobagyi GN et al: Phase II evaluation of Interferon added to Tamoxifen in the treatment of metastatic breast cancer. Breast Cancer Res Treat 1991 (18):165-170

42 Buzzi F, Brugia M, Rossi G, Giustini L et al: Combination of beta-Interferon and Tamoxifen as a new way to overcome clinical resistance to Tamoxifen in advanced breast cancer. Anticancer Res 1992 (12):869-871

43 Miglietta L, Repetto L, Gardin G et al: Tamoxifen and alpha-Interferon in advanced breast cancer. J Chemother 1991 (3):383-386

44 Vanderberg T, Skillings J: Phase II study of Interferon alpha and Tamoxifen in patients with metastatic breast cancer previously unresponsive to Tamoxifen. Proc Ann Meet Am Assoc Cancer Res 1992 (33):A1274

45 Repetto L, Gardin G, Campora E et al: Tamoxifen and Interferon beta in the treatment of metastatic breast cancer: a phase II study. Ann Oncol 1992 (3-suppl. 5):297-302

46 Rechia F, Marchionni F, Rabitti G: Phase II trial of Tamoxifen, beta Interferon and retinoids in metastatic breast cancer. Proc Ann Meet Am Soc Clin Oncol 1991 (10):A88

47 Pevetz T, Kaplan A, Bavak V, Baider L, Sulkes A, Rentschler F, Stephanos S, Catane R: Interferon as a modulation of hormonal therapy in breast cancer: preliminary results of a randomized study. Proc Ann Meet Am Assoc Cancer Res 1992 (33):A2000

48 Spielmann M, Kayitalire L, Boussen H et al: Salvage therapy with Fluorouracil, folinic acid in continuous infusion and SC Interferon alpha-2A in metastatic breast cancer: a preliminary report. Proc Ann Meet Am Soc Clin Oncol 1992 (11):A124

49 Walters RS, Esparza L, Theriault RL et al: The impact of the addition of recombinant alpha Interferon to 5-Fluorouracil in patients with metastatic breast cancer. Proc Ann Meet Am Soc Clin Oncol 1991 (10):A107

50 Pronzato P, Bertelli G, Amoroso D et al: Cisplatin and recombinant alpha Interferon in advanced breast cancer. Ann. Oncol. 1990 (12-2):150-151

51 Scnalke D, Wildfaug I, Grundel O et al: A pilot trial of 5 Fluorouracil, Leucovorin and R IFN-alpha in refractory metastatic breast cancer. Ann Oncol 1992 (3-suppl 15):34-36

52 Walters R, Parkinson D, Talpaz M et al: A phase II trial of recombinant Interleukin-2 and recombinant alpha Interferon in metastatic breast cancer. Proc Ann Meet Am Soc Clin. Oncol 1990 (9):A134

53 Fentiman IS, Balkwill FR, Cuzick J, Hayward JL et al: A trial of human alpha Interferon as an adjuvant agent in breast cancer after loco-regional recurrence. Eur J Surg Oncol 1987 (13-5):425-428

54 Buzdar A, Hortobagyi G, Kau S, Holmes F, Fraschini G, Ames F, Singletary S, Gutterman J: Escalating dose-intensive adjuvant therapy with Doxorubicin and Cyclophosphamide in stage II or III breast cancer with or without alpha-Interferon: a prospective randomized trial. Proc Ann Meet Am Soc Clin Oncol 1989 (8):A106

55 Early Breast Cancer Trialists' Collaborative Group: Systemic treatment of early breast cancer by hormonal, cytotoxic, or immune therapy. Lancet 1992 (339):1-15, 71-85

Renal Cell Carcinoma

P.A. Ruffini and C. Gambacorti-Passerini

Division of Experimental Oncology D, Istituto Nazionale Tumori, Via Venezian 1, 20133 Milan, Italy

Metastatic renal cell carcinoma (RCC) is at present an incurable disease. The median survival time for patients with metastatic RCC is approximately 10 months with less than 10% of patients surviving for 2 years [1]. This poor prognosis depends on the fact that conventional therapy is ineffective, with an objective response rate of less than 20% for hormonal and chemotherapeutic agents given either alone or in combination [2]. These disappointing results, the recent availability of recombinant cytokines and the fact that spontaneous regressions of RCC (presumably mediated by the host immune response) have been reported, have spawned interest in the immunobiology of RCC with the aim of understanding the immunological relationship between the host and the tumour, and evaluating whether immunotherapy could represent a novel approach to this malignancy. Interferons (IFNs), in particular, have been investigated in the treatment of advanced RCC either as a single agent or in combination with cytotoxic drugs or other cytokines.

Preclinical Aspects

IFNs have been investigated as potential anticancer agents because of their ability to activate the host immune system and their direct antiproliferative effect on tumour cells.

It is now well established that if tumour-specific neo-antigens are expressed by human tumour cells, the concomitant expression of major histocompatibility complex (MHC) class I and II molecules is needed to make them accessible to MHC-restricted T lymphocytes. This could enable the host immune system to generate a cell-mediated antitumour response *in vivo*. IFNs exert their multiple activities primarily by inducing and/or amplifying the expression of a large variety of genes (together with the proteins encoded), including those of MHC. IFN-α and γ are able to upregulate the expression of class I and II MHC molecules on the surface of cancer cells, although IFN-α is less active than IFN-γ in modulating class II MHC antigens. This has been demonstrated also in the case of RCC cell lines [3]. Moreover, a correlation between MHC molecules inducibility by IFNs and sensitivity to IFN therapy has been shown in human renal tumour xenografts in nude mice [3]. It should be noted, however, that in this model T-cell-mediated effector mechanisms cannot explain the results observed. In patients, it has been found that the pretreatment human leukocyte antigen (HLA) phenotype of RCC cells has no predictive value for outcome of IFN immunotherapy [4], even though a role for treatment-induced changes in HLA expression *in vivo* cannot be excluded.

For a specific immune response against tumour cells, antigen-independent interactions between adhesion molecules are also required in addition to antigen-specific recognition between T lymphocytes and target cells. One of these adhesion molecules, intercellular adhesion molecule-1 (ICAM-1), has been shown to play an important role in tumour cell specific lysis by lymphocytes. Recently, the expression of ICAM-1 on RCC cell lines has been investigated [5]. This study revealed a frequent expression of ICAM-1 on RCC cell lines (23 positive cases among 28 primary RCC) and a significant correlation between ICAM-1 expression and the degree of mononuclear cell infiltration (mainly T lymphocytes) in the tumour. In the light of these data the observation that ICAM-1 expression on RCC cells is highly susceptible to IFN-γ treatment [5] could be significant for the host antitumour response, although the evaluation of ICAM-1 expression in metastatic RCC was not carried out.

In conclusion, the ability of IFNs to upregulate the expression of MHC class I and II molecules and cell adhesion molecules on RCC cells can enhance both the antigen presenting capacity of tumour cells and their susceptibility to lysis by MHC-restricted cytotoxic T lymphocytes. It should be noted, however, that no clear evidence has been obtained yet that RCC express antigens that, as in the case of melanoma, can be recognised by T cells in a MHC-restricted fashion [6,7], thus casting doubts that even the up-regulation of MHC antigens and adhesion molecules can be of any help in improving such a recognition.

It is well known that IFNs can mediate a direct antiproliferative effect on some types of tumour cells. This has been demonstrated also for RCC cells. Of 16 RCC cell lines tested for intrinsic sensitivity to the growth inhibitory effect of IFN-α, 7 showed a marked growth inhibition with concentrations of IFN-α which were comparable to those achieved in patients with RCC after a standard IFN-α protocol (i.e., 100 to 1,000 IU/mL) [8]. It should be noted that the RCC cell lines tested were from either primary or metastatic sites and 6 out of 11 lines obtained from metastases were shown to be sensitive to IFN-α. These data were confirmed when the same set of RCC cell lines was grown not in tissue cultures but as tumour xenografts in mice and their proliferation was also markedly inhibited by systemically administered IFN-α [8].

One of the mechanisms by which IFN-α can affect RCC cell proliferation may be the downregulation of epidermal growth factor receptor (EGFR) expression on RCC cells. EGF stimulates proliferation of both IFN-sensitive and resistant RCC cell lines. It has been shown that IFN-α is able to block the EGF-stimulated proliferation of RCC cells by inhibiting EGFR synthesis only in RCC cell lines sensitive to IFN-α while in RCC cell lines resistant to the antiproliferative action of IFN-α, IFN-α treatment has no effect on EGFR expression [9].

Recently, fresh RCC cells and cell lines have been shown to express interleukin-6 (IL-6) mRNA and to produce biologically active IL-6 in the supernatants [10]. IL-6 receptor has been detected on RCC cells and anti-IL-6 antibodies block the proliferation of RCC cells in vitro, thus indicating that IL-6 stimulates RCC growth in an autocrine fashion [10]. Consistent with these observations is the finding that IL-6 levels are increased in the serum of patients with metastatic RCC as compared to normal donors [11]. IFN-γ has been shown in vitro to interrupt the autocrine IL-6-mediated loop generated by RCC cells leading to tumour growth inhibition [12]. These authors found that IFN-γ treatment inhibited IL-6 release by RCC cells in a dose-dependent manner without affecting IL-6 receptor expression on RCC cells; addition of exogenous IL-6 to IFN-γ-treated cells restored a normal proliferative potential. Although further investigations are needed, these results suggest that IFN-γ may exert an inhibitory effect on RCC growth by this mechanism.

Several observations in preclinical systems have indicated that another important function of IFNs as anticancer agents may lie in their capacity to modulate the activity of other cytotoxic or biological agents. The ability of IFNs to synergistically potentiate the activity of a wide variety of chemotherapeutic drugs against human and experimental tumours has been clearly demonstrated both in vitro and in animal models [13,14]. Unfortunately, the results of clinical trials using combinations of IFNs with chemotherapeutic agents have generally been disappointing even in the case of RCC, where the addition of cytotoxic drugs to IFN-α appeared to offer no therapeutic advantage [15].

Combinations of IFNs and other biological agents have also been investigated. The rationale for combining IFNs and interleukin-2 (IL-2) is that either agent alone has antitumour activity against RCC [2] and that combination of these two agents shows a synergistic antitumour effect in the mouse model [16,17], even though the mechanisms by which IFNs and IL-2 mediate tumour regression are not yet clear. It has been postulated, however, that IFNs enhance the expression on the tumour cells of MHC, adhesion molecules and/or other molecules, thus rendering such cells more susceptible to the IL-2-induced lymphocyte cytotoxicity.

Expression of interleukin-4 (IL-4) receptors has been demonstrated on RCC cells [18]. IL-4 has been shown to inhibit in vitro growth of RCC cell lines. Combination of IL-4 and IFN-γ produced a significant augmentation of the inhibitory effect on RCC cell proliferation [18].

The combination of IFN-α and γ has also been studied. The *in vitro* antitumour activity of such a combination has been tested on colony formation of two human RCC xenografts [19]. A synergistic effect on colony formation of both tumour lines was detected, thus suggesting that IFN-α and γ can exert a direct, synergistic antiproliferative effect on RCC cells. The clinical activity of IFN-α and γ combination has been demonstrated in a murine renal tumour model [20]. In this model, the potentiation of a T-cell-mediated mechanism is probably responsible for the therapeutic effects. In fact, euthymic mice successfully treated with IFN combination were found to be immune to rechallenge with the same tumour line while no significant therapeutic advantage in terms of survival was observed in nude mice. These results suggest that in this model the ability of IFN-α and γ combination to induce rejection of the intraperitoneally injected renal tumour depended more upon activation of T-cell-mediated immunity than upon direct inhibitory activity on tumour growth [20]. As pointed out above, however, the evidence for such a T-cell-mediated reaction to RCC in humans remains questionable. The rationale for α and γ IFN combination depends upon the fact that IFN-α and γ bind to different cell surface receptors and IFN-γ can induce the expression of cell surface receptors to IFN-α [21].

Finally, the recent development of techniques of retroviral-mediated gene transfer offers the possibility of generating more immunogenic tumour cell vaccines. RCC cells genetically engineered to produce IFN-γ or IL-2 have been generated [22] and a clinical protocol has been approved for the use of such cells for the *in vivo* immunisation of patients with metastatic RCC.

The preclinical results obtained in both *in vitro* and animal systems are summarised in Table 1.

Clinical Aspects

Among the various biological response modifiers, IFNs have been the most extensively investigated agents in patients with metastatic RCC with more than 1,000 patients evaluated in numerous clinical trials. In 1983, the University of California-Los Angeles (UCLA) [23] and the M.D. Anderson group [24], in independent studies, reported on the regression of metastatic RCC with partially purified human leukocyte IFN. Objective responses (complete response, CR, and partial response, PR) occurred in 16.5% and 26% of patients, respectively. Following these initial results, a number of clinical trials with IFNs have confirmed a reproducible objective response rate of 15% to 20% (Table 2).

However, the median duration of response is approximately 6 months, and 3 out of 4 responding patients achieve only PR.

Toxicity in these trials included symptoms commonly associated with IFNs, with a flu-like syndrome consisting of malaise, fever, chills or muscle aches occurring in almost all patients. Such side effects are dose related but are rarely severe at a daily dosage of less than 10 million units (MU). An optimal dosage and schedule have not been determined, but a dose of 5-10 MU given at least 3 times per week appears to result in the best therapeutic index. The route of administration has not been associated with response. Responses seem to correlate with some pretreatment characteristics, i.e., prior nephrectomy, good performance status and lung as predominant site of disease [38,39].

Since the cumulative clinical experience has revealed that only a subset of patients with advanced RCC respond to IFN therapy, and given the toxicity IFN therapy involves, a means of predicting which patients will respond would be very useful. The only marker predicting a response to IFN-α in RCC is a kidney-associated differentiation glycoprotein named gp160 [8]. The expression of this molecule on tumour lines from either primary or metastatic RCC was found to correlate with resistance to the antiproliferative effect of IFN-α, while RCC cell lines lacking expression of gp160 were shown to be sensitive to the same cytokine. The correlation of IFN-α-sensitive or resistant phenotype with gp160 expression was demonstrated both *in vitro* and *in vivo* using a mouse model [8]. If this finding can be applied in a clinical setting, then the absence of gp160 expression in a biopsied renal tumour may identify those patients with a high probability of achieving a positive clinical response to IFN-

Table 1. Biomodulation of renal cell carcinoma by interferons: preclinical studies

Type of IFN	Experimental system	Biological effect	Clinical significance	Reference
α,γ	Human RCC xenografts in nude mice	Up-regulation of HLA class I and II molecules on RCC cells	Could make RCC cells accessible to HLA-restricted (i.e. specific) anti-RCC lymphocytes	[3]
γ	Human RCC cell lines cultured *in vitro*	Up-regulation of ICAM-1 expression on RCC cells	Elevation of ICAM-1 expression on RCC might increase the host immune reaction	[5]
α	Human RCC cell lines cultured *in vitro* and transplanted in nude mice	Direct inhibition of proliferation *in vitro* and *in vivo*	IFN treatment may delay the *in vivo* growth of at least a subset of RCC	[8]
α	Human RCC cell lines cultured *in vitro*	Block of EGF-stimulated RCC cell proliferation through inhibition of EGFR synthesis by tumour cells	IFN treatment may block RCC growth *in vivo* by inhibiting EGFR expression on RCC cells	[9]
γ	Human RCC cell lines cultured *in vitro*	Inhibition of RCC cell proliferation through interruption of IL-6-mediated autocrine growth stimulation	IFN might reduce RCC growth *in vivo* by interrupting the autocrine IL-6-mediated loop	[12]
α+γ	Human RCC xenografts	Synergistic inhibition of tumour growth *in vitro*	IFN-α and γ combination may exert a direct, synergistic inhibitory effect on RCC growth	[19]
α+γ	Murine renal tumour	Successful treatment resulting in systemic antitumour immunity	IFN-α + γ treatment may elucidate a T-cell-mediated antitumour response	[20]
γ	Human RCC cells transduced *in vitro* with IFN-γ gene	Augmented expression of HLA class-I and II molecules and ICAM-1	Use as live tumour cell vaccine for *in vivo* immunisation of RCC patients	[22]

α therapy.

Unfortunately, even if sustained clinical responses have been reported, a recent study comparing the therapeutic efficacy and, more important, the impact on survival of IFN-α vs hormonal agents, chemotherapy and immunotherapy other than IFN in patients with RCC, demonstrated that even if IFN-α was the only agent which produced objective tumour regressions, no difference in overall survival could be demonstrated in these groups [31]. With regard to the immunomodulatory effects of IFNs in RCC patients, a group recently reported on immune changes observed in patients with RCC treated with IFN-α [32]. All the 12 immunologically evaluable patients had reduced responses compared with normal donors in any of the immunological parameters investi-

Table 2. Interferons as single agent therapy in metastatic renal cell carcinoma: clinical studies

Type of IFN	Number of evaluable patients	CR + PR	%PR	Median duration of response	Reference
Cantell	43	1 + 6	16.5	Not eval.	[23]
Cantell	19	0 + 5	26	6+ months	[24]
Cantell	30	1 + 2	10	Not eval.	[25]
α-2a lymphoblastoid [1]	226	4 + 36	18	79 days	[26]
α–2b	20	0 + 1	5	15 months	[27]
Cantell	43	- 2	14	Not eval.	[28]
α-2a	19	- 2	26	Not eval.	[28]
α-2a	19	1 + 4	26	283 days	[29]
Lymphoblastoid	25	2 + 4	24	6.5 ± 6.4 months [3]	[30]
α-2a, lymphoblastoid [4]	22	0 + 2	9	Not eval.	[31]
α-2b	26	0 + 4	15	Not eval.	[32]
β	25	1 + 4	20	225 days	[33]
γ	13	0 + 0	0	0	[34]
γ	41	1 + 3	10	6, 9 months [5]	[35]
γ	20	2 + 4	30	14+ months	[36]
γ + α	30	2 + 6	27	Not eval.	[37]

[1] 153 patients received IFN α-2a and 73 were treated with human lymphoblastoid IFN; [2] The authors did not present the number of CR and PR obtained; [3] Mean duration of response; [4] 11 patients received IFN α-2a and 11 were treated with human lymphoblastoid IFN; [5] For CR and PR, respectively

gated (i.e., T-lymphocyte proliferation in autologous and allogeneic mixed lymphocyte reaction (MLR), IL-2 production, expression of IL-2 receptors during the allogeneic MLR, and interleukin-1 (IL-1) production by peripheral blood monocytes) before IFN-α treatment. During therapy these deficient immune responses were restored to almost normal levels in 5 out of 12 patients (4 responders and 1 with stable disease) whereas no significant changes were observed in the other 7 patients (all non-responders). The authors conclude that there was a good correlation between the changes observed in the immunological parameters investigated and the clinical outcome of these patients [32].

In order to evaluate the IFN-induced stimulation of the host immune system in patients with metastatic RCC, peripheral blood mononuclear cells (PBMC) of patients were tested for natural killer (NK) and lymphokine-activated killer (LAK) activity both before and after IFN-α therapy. A slight but not statistically significant increase in NK cell cytotoxicity occurred in comparison to the one observed before the onset of immunotherapy while no LAK activity could be detected [40].

The demonstration in experimental models of synergy between IFNs and IL-2 was tested in clinical trials in RCC. Initial results utilising combination of IL-2 and IFNs suggest no therapeutic advantage as compared to single-agent IFN [38]. In addition, toxicity for IFNs/IL-2 combination has generally exceeded that of IFN alone. Atzpodien and co-workers, however, reported a 29% response rate among 34 patients treated in an outpatient setting, with mild toxicity [41]. Clinical responses in this study were associated with a mean peripheral blood eosinophil count of more than 1,000/μl. Similar clinical results have been reported by other groups [42], and one of them recently reported their clinical and immunological results using an IFN-α/IL-2 combination [43]. Of the 6 patients evaluated, none had a therapeutic response. Of interest, however, were the findings that, although an increase in overall numbers of LAK and NK cells followed each infusion cycle, the antitumour activity of these cells as measured by *in vitro* assays, as well as the secondary cytokine release, actually reached a plateau and even decreased after two cycles of therapy.

The simultaneous administration of IFN-α and γ showed in preliminary trials no therapeutic advantage vs. IFN alone and caused an increased toxicity [44]. In a recent report [37], IFN-γ and α administered sequentially induced a clinical response in 8 out of 30 evaluable patients. The results of this study suggest that the efficacy and toxicity profile associated with combination IFN therapy may be improved by administering these agents sequentially, as opposed to simultaneously. Several immune parameters have been analysed in these patients, including changes in T-cell and NK cell population, NK activity and 2',5' oligoadenylate synthetase (2'5'S) levels [45]. Pre- and post-treatment comparisons for the 28 immunologically evaluable patients showed significant decrease in peripheral blood NK cell activity, CD8+, CD56+ and CD16+ cells. Significant increases were seen in the percentage of CD4+ cells, the helper/ suppressor (CD4/CD8) ratio and 2'5'S. The decrease in the percentage of CD8+ cells was the only factor found in association with positive clinical outcome. This finding suggests that a T-cell-mediated antitumour mechanism may be involved and is consistent with the results obtained in the mouse model [20].

Conclusions

There is no satisfactory therapy for patients with metastatic RCC. Immunotherapy remains the mainstay of non-surgical treatment and IFNs have provided a reproducible response rate of 15% to 20%. The antitumour efficacy of IFNs could relate to: a) effects on the tumour (i.e., direct growth inhibition, augmentation of tumour immunogenicity); b) stimulation of immune response to the tumour; c) a combination of effects on the tumour and stimulation of the host immune response. Many experimental and clinical data indicate that IFNs can exert both a direct antiproliferative effect against RCC cells and can potentiate the host antitumour reaction. Demonstration that this direct/indirect activity of IFNs is selective for tumour cells is still lacking. The analysis of this issue, as it has been addressed for antimelanoma lymphocytes [46,47], is of paramount importance and should be the focus of future studies.

From a clinical point of view, the major problem to be solved is the identification of possible responders to IFN treatment. Even though both preclinical and clinical characteristics have been investigated as predictors of a positive clinical outcome, to date we still lack a basis for identifying those patients who will benefit from IFN immunotherapy.

REFERENCES

1   deKernion JB, Ramming KP, Smith RB: The natural history of metastatic renal cell carcinoma: a computer analysis. J Urol 1978 (120):148-152

2   Reese JH: Renal cell carcinoma. Curr Opin Oncol 1992 (4):427-434

3   Beniers AJMC, Peelen WP, Debruyne FMJ, Schalken JA: HLA-class-I and -class-II expression on renal tumor xenografts and the relation to sensitivity for IFN-α, IFN-γ and TNF. Int J Cancer 1991 (48):709-716

4   Mattijssen V, Van Moorselaar J, De Mulder PH, Schalkwijk L, Ruiter DJ: Human leucocyte antigen expression in renal cell carcinoma lesions does not predict the response to interferon therapy. J Immunother 1992 (12):64-69

5   Tomita Y, Nishiyama T, Watanabe H, Fujiwara M, Sato S: Expression of intercellular adhesion molecule-1 (ICAM-1) on renal-cell cancer: possible significance in host immune responses. Int J Cancer 1990 (46):1001-1006

6   Belldegrun A, Muul LM, Rosenberg SA: Interleukin-2 expanded tumor-infiltrating lymphocytes in human renal cell cancer: isolation, characterization and antitumor activity. Cancer Res 1988 (48):206-214

7   Finke JH, Rayman P, Alexander J, Edinger M, Tubbs RR, Connelly R, Pontes E, Bukowski R: Characterization of the cytolytic activity of CD4+ and CD8+ tumor-infiltrating lymphocytes in human renal cell carcinoma. Cancer Res 1990 (50):2363-2370

8   Nanus DM, Pfeffer LM, Bander NH, Bahri S, Albino AP: Antiproliferative and antitumor effects of α-interferon in renal cell carcinomas: correlation with the expression of a kidney-associated differentiation glycoprotein. Cancer Res 1990 (50):4190-4194

9   Eisenkraft BL, Nanus DM, Albino AP, Pfeffer LM: α-interferon down-regulates epidermal growth factor receptors on renal carcinoma cells: relation to cellular responsiveness to the antiproliferative action of α-interferon. Cancer Res 1991 (51):5881-5887

10 Miki S, Iwano M, Miki Y, Yamamoto M, Tang B, Yokokawa K, Sonoda T, Hirano T, Kishimoto T: IL-6 functions as an autocrine growth factor in renal cell carcinoma. FEBS Lett 1989 (250):607-610

11 Blay JY, Negrier S, Combaret V, Attali S, Goillot E, Merrouche Y, Mercatello A, Rovault A, Tourani JM, Moskovtchenko JF, Philip T: Serum level of IL-6 as a prognosis factor in metastatic renal cell carcinoma. Cancer Res 1992 (52):3317-3322

12 Gruss HJ, Brach MA, Mertelsmann RH, Herrmann F: Interferon-γ interrupts autocrine growth mediated by endogenous IL-6 in renal cell carcinoma. Int J Cancer 1991 (49):770-773

13 Wadler S, Schwartz EL: Antineoplastic activity of the combination of interferon and cytotoxic agents against experimental and human malignancies: a review. Cancer Res 1990 (50):3473-3486

14 Parmiani G, Rivoltini L: Biologic agents as modifiers of chemotherapeutic effects. Curr Opin Oncol 1991 (3):1078-1086

15 Neidhart JA, Anderson SA, Harris JE, Rinehart JJ, Laszlo J, Dexeus FH, Einhorn LH, Trump DL, Benedetto PW, Tuttle RL, Smalley RV: Vinblastine fails to improve response of renal cancer to interferon-α-n1: high response rate in patients with pulmonary metastases. J Clin Oncol 1991 (9):832-837

16 Brunda MJ, Bellantoni D, Sulich V: In vivo anti-tumor activity of combinations of IFN α and interleukin-2 in a murine model. Correlation of efficacy with the induction of cytotoxic cells resembling natural killer cells. Int J Cancer 1987 (40):365-371

17 Cameron RB, McIntosh JK, Rosenberg SA: Synergistic antitumor effects of combination immunotherapy with recombinant interleukin-2 and a recombinant hybrid α-interferon in the treatment of established murine hepatic metastases. Cancer Res 1988 (48):5810-5817

18 Hoon DS, Okun E, Banez M, Irie RF, Morton DL: Interleukin-4 alone and with γ-interferon or α-tumor necrosis factor inhibits cell growth and modulates cell surface antigens on human renal cell carcinoma. Cancer Res 1991 (51):5687-5693

19 Beniers AJMC, Peelen WP, Hendriks BT, Schalken JA, Debruyne FMJ: Effect of α- and γ-interferon and tumor necrosis factor on colony formation of two human renal tumor xenografts in vitro. Sem Surg Oncol 1988 (4):195-198

20 Sayers TJ, Wiltrout TA, McCormick K, Husted C, Wiltrout RH: Antitumor effects of α-interferon and γ-interferon on a murine renal cancer (Renca) in vitro and in vivo. Cancer Res 1990 (50):5414-5420

21 Hannigan GE, Fish EN, Williams BRG. Modulation of human interferon α receptor expression by human interferon γ. J Biol Chem 1984 (259):8084-8086

22 Gastl G, Finstad CL, Guarini A, Bosl G, Gilboa E, Bander NH, Gansbacher B: Retroviral vector-mediated lymphokine gene transfer into human renal cancer cells. Cancer Res 1992 (52):6229-6236

23 deKernion JB, Sarna G, Figlin RA, Lindner A, Smith RB: The treatment of renal cell carcinoma with human leucocyte α-interferon. J Urol 1983 (130):1063-1066

24 Quesada JR, Swanson DA, Trindade A, Gutterman JU: Renal cell carcinoma: antitumor effects of leucocyte interferon. Cancer Res 1983 (43):940-947

25 Kirkwood JM, Harris JE, Vera R, Sandler S, Fischer DS, Khaudekar J, Ernstoff MS, Gordon L, Lutes R, Bonomi P, Lytton B, Cobleigh M, Taylor IV SJ: A randomized study of low and high doses of leukocyte α-interferon in metastatic renal cell carcinoma: the American Cancer Society collaborative trial. Cancer Res 1985 (45):863-873

26 Umeda T, Niijima T: Phase II study of α-interferon on renal cell carcinoma. Cancer 1986 (58):1231-1235

27 Kempf RA, Grunberg SM, Daniels JR, Skinner DG, Venturi CL, Spiegel R, Neri R, Greiner JM, Rudnick S, Mitchell MS: Recombinant interferon α-2 (INTRON A) in a phase II study of renal cell carcinoma. J Biol Resp Modif 1986 (5):27-35

28 Sarna G, Figlin R, deKernion J: Interferon in renal cell carcinoma. The UCLA experience. Cancer 1987 (59):610-612

29 Figlin RA, deKernion JB, Mukamel E, Palleroni AV, Itri LM, Sarna GP: Recombinant Interferon α-2a in

metastatic renal cell carcinoma: assessment of antitumor activity and anti-Interferon antibody formation. J Clin Oncol 1988 (6):1604-1610

30 Fujita T, Asano H, Naide Y, Ono Y, Ohshima S, Suzuki K, Aso Y, Ariyoshi Y, Fukushima M, Ota K: Antitumor effects of human lymphoblastoid interferon on advanced renal cell carcinoma. J Urol 1988 (139):256-258

31 Retsas S, Bafaloukos D, Brunt AM: α-interferons: impact on survival of patients with renal cell carcinoma. Clin Oncol 1991 (3):273-277

32 Kosmidis PA, Baxevanis CN, Tsavaris N, Papanastasiou M, Anastasopoulos E, Bacoyiannis C, Mylonakis N, Karvounis N, Bafaloukos D, Karabelis A, Papamichail M: The prognostic significance of immune changes in patients with RCC treated with interferon-α2b. J Clin Oncol 1992 (10):1153-1157

33 Kinney P, Triozzi P, Young D, Drago J, Behrens B, Wise H, Rinehart JJ: Phase II trial of interferon-beta-serine in metastatic renal cell carcinoma. J Clin Oncol 1990 (8):881-885

34 Rinehart JJ, Malspies L, Young D, Neidhart JA: Phase I/II trial of human recombinant interferon γ in renal cell carcinoma. J Biol Resp Modif 1986 (5):300-308

35 Garnick MB, Reich SD, Maxwell B, Coval-Goldsmith S, Richie JP, Rudnick SA: Phase I/II study of recombinant interferon γ in advanced renal cell carcinoma. J Urol 1988 (139):251-255

36 Aulitzky W, Gastl G, Aulitzky WE, Herold M, Kemmler J, Mull B, Frick J, Huber C: Successful treatment of metastatic renal cell carcinoma with a biologically active dose of recombinant interferon-γ. J Clin Oncol 1989 (7):1875-1884

37 Ernstoff MS, Nair S, Bahnson RR, Miketic LM, Banner B, Gooding W, Day R, Whiteside T, Hakala T, Kirkwood JM: A phase IA trial of sequential administration recombinant DNA-produced interferons: combination recombinant interferon γ and recombinant interferon α in patients with metastatic renal cell carcinoma. J Clin Oncol 1990 (8):1637-1649

38 Muss HB: The use of interferon in renal cell carcinoma. Eur J Cancer 1991 (27,S):84-87

39 Wadler S: The role of interferons in the treatment of solid tumors. Cancer 1992 (70,S):949-958

40 Feruglio C, Zambello R, Trentin L, Bulian P, Franceschi T, Cetto GL, Semenzato G: Cytotoxic in vitro function in patients with metastatic RCC before and after α-2b-interferon therapy. Cancer 1992 (69):2525-2531

41 Atzpodien J, Poliwoda H, Kirchner H: α-interferon and interleukin-2 in renal cell carcinoma: studies in nonhospitalized patients. Sem Oncol 1991 (18,S):108-112

42 Figlin RA, Belldegrun A, Moldawer N, Zeffren J, deKernion J: Concomitant administration of recombinant human interleukin-2 and recombinant interferon α-2a: an active outpatient regimen in metastatic renal cell carcinoma. J Clin Oncol 1992 (10):414-421

43 Pichert G, Jost LM, Fierz W, Stahel RA: Clinical and immune modulatory effects of alternative weekly interleukin-2 and IFN-α-2a in patients with advanced renal cell carcinoma and melanoma. Br J Cancer 1991 (63):287-292

44 Quesada JR, Evans L, Saks SR, Gutterman JU: Recombinant interferon α and γ in combination as treatment for metastatic renal cell carcinoma. J Biol Resp Modif 1988 (7):234-239

45 Ernstoff MS, Gooding W, Nair S, Bahnson RR, Miketic LM, Banner B, Day R, Whiteside T, Titus-Ernstoff L, Kirkwood JM: Immunological effects of treatment with sequential administration of recombinant interferon γ and α in patients with metastatic renal cell carcinoma during a phase I trial. Cancer Res 1992 (52):851-856

46 Parmiani G, Anichini A, Fossati G: Cellular immune response against autologous human malignant melanoma: are in vitro studies providing a framework for a more effective immunotherapy? JNC 1990 (82):361-370

47 Anichini A, Maccalli C, Mortarini R, Salvi S, Mazzocchi A, Squarcina P, Herlyn M, Parmiani G: Melanoma cells and normal melanocytes share antigens recognized by HLA-A2 restricted cytotoxic T cell clones from melanoma patients. J Exp Med 1993 (177):989-998

Prostate Cancer

K. Pummer[1], P. Pürstner[2], G. Lanzer[3], U. Pätzold[3] and H.L. Auner[2]

1   Universitätsklinik für Urologie, Karl Franzens Universität, Graz
2   Universitätsklinik für Gynäkologie und Geburtshilfe, Karl Franzens Universität, Graz
3   Department für Transfusionsmedizin und Immunhämatologie der Chirurgischen Universitätsklinik, Karl Franzens Universität, Graz, Austria

Adenocarcinoma of the prostate is one of the most common visceral cancers in males. It was estimated that in 1992 132,000 new cases were diagnosed in the United States and 34,000 prostatic cancer deaths occurred [1]. At the time of presentation, more than 50% of newly diagnosed patients either have locally advanced or metastatic disease. It is also widely recognised, however, that the prevalence of prostatic cancer found at autopsy largely surpasses the prevalence of clinically detectable disease [2]. In simple terms, there is a 14-fold greater risk of developing prostate cancer than there is of dying from it [3].

The possible explanation for this apparent discrepancy is that, on the one hand, there is a relatively vast reservoir of biologically quiescent or latent lesions of which only a few progress to become clinically manifest, and, on the other hand, prostatic cancer is characteristically heterogeneous in both its natural history and its malignant potential. The natural history, which is the end result of interactions between host, tumour, and environmental factors influencing the course of disease, and the clinical behaviour remain obscure and ambiguous, which may be attributed to both our present ignorance of the biological behaviour of this tumour and the lack of well-controlled, unbiased and randomised studies comparing treated and untreated patients with similar tumours and matched by age, grade, stage and even race [4].

Although prostate cancer confined to the prostate gland at diagnosis is curable with radical prostatectomy or even radiation therapy, the mean survival for patients with metastatic disease is less than 2 years [5]. Treatment for these advanced cases is based on the fact that prostatic cancer cell growth is hormone dependent. The era of hormonal manipulation was ushered in by Huggins and Hodges in 1941, who succeeded in demonstrating the beneficial effect of hormonal alteration with either oestrogens or bilateral castration on the disease process [6]. Since that time, a variety of new pharmacological agents capable of interrupting the hypothalamic-pituitary-testicular axis to eliminate, reduce, or otherwise block the action of testosterone, the major androgenic hormone, have been developed.

Early in the 1980s, the concept of combining an analogue of gonadotropin-releasing hormone or orchidectomy with an antiandrogen has been developed in the attempt to maximise the effects of androgen ablation [7]. This approach is primarily based on the hypothesis that progression of tumour growth after conventional endocrine therapy may result from inadequate suppression of androgens of adrenal origin, because it has been observed that after gonadal ablation, intracellular levels of dihydrotestosterone in prostatic cells remain disproportionately high, despite decreases in serum testosterone to castration levels [7]. Although prolonged progression-free survival and overall survival have been reported, the improvement is small but important, and is most pronounced in patients with minimal disease [8]. Despite some progress in hormonal manipulation, relapse following initial successful treatment is inevitable and the possibilities of further successful treatment are limited. The reason for progression is that prostatic tumour cells are heterogeneous at least with regard to their androgen requirements for growth, being composed of preexisting clones of androgen-dependent and independent tumour cells. Hormonal ablation in such a context does not affect the continuous growth of androgen-independent cells [9]. To overcome this situation, chemotherapy has been employed. However, despite beneficial evidence in animal models and encouraging preliminary clinical results, substantial advances failed to occur [9,10].

In view of this therapeutic dilemma, there is a need for the development of new approaches to the treatment of prostatic cancer. In the 1960s, it has been reported that interferons have antiproliferative effects in different experimental models [11]. Their possible role in the treatment of prostatic cancer derives from a variety of cellular and molecular changes that reportedly occur in association with interferon. However, the clinical effectiveness of interferons in prostatic cancer is a subject of controverse and remains to be established yet.

In this chapter we try to give an overview of our current understanding of interferons' action in prostate cancer. In this context, the current knowledge of changes in hormone receptor levels, direct antiproliferative properties, and immunological changes will be presented, followed by a review of clinical trials which have been performed so far.

Interferons and Androgen Receptors

The molecular mechanisms of growth of prostate cancer as well as normal prostate development are not completely understood. One factor that clearly plays an important role in growth regulation is the androgen receptor. Currently, both cytoplasmic and nuclear androgen receptors are known. While cytoplasmic androgen-receptor concentrations are not related to the clinical outcome of hormonal manipulation, the nuclear androgen-receptor concentrations were found to be significantly higher in a group of patients responding to hormonal therapy. This suggests that the hormone-dependence of tissues can be predicted by the presence of at least partially normal functioning specific receptors for particular hormones [12]. Unresponsiveness of tumour growth might be correlated with the absence of the receptor or the presence of a mutated receptor. The andro-

gen receptor and androgen receptor mRNA could be detected even in the androgen-independent Dunning sublines, which were directly derived from the original androgen-dependent tumour. This indicates that the loss of androgen receptor expression might be a functionally late step in progressive tumour growth. The possibility that loss of steroid receptor may be a consequence rather than a cause of insensitivity has also been discussed, because it has been observed that culture of steroid-sensitive cell lines using prolonged withdrawal of steroids generated cells that were insensitive in terms of proliferation but retained functional receptors [13,14]. Thus, knowledge of the mechanism of androgen-independent growth of prostatic cancer is still limited and data have been obtained showing that androgen receptor levels can be high in the prostatic tissue of patients with progressive disease, even years after orchiectomy [15].

Differences in receptor content between primary tumours, usually obtained at early stages of the disease, and metastatic sites, considered to represent a dedifferentiated, more aggressive cell population, have rarely been investigated, although different hormonal responsiveness might be expected. In a recent study, Ekman and Brolin reported that the total androgen-receptor content was similar in metastatic and primary lesions, however, with a shift towards a cytoplasmic predominance in metastatic tissue obtained from lymph nodes [16]. They speculated that low amounts of nuclear receptor may reflect tumour tissue less susceptible to endocrine manipulations.

Interferon has been shown to enhance the steroid hormone-receptor level in different *in vitro* models and to sensitise breast cancer cells to the antiproliferative effects of the antioestrogen tamoxifen, which knowingly acts via oestrogen receptors [17-20]. On the basis of these findings, it can be expected that exposure to interferon can make hormone-unresponsive prostatic tumour cells responsive to both dihydrotestosterone (DHT) and, in turn, to hydroxyflutamide, which might have important clinical implications in the treatment of progressive cancer. In this context, Sica et al. investigated androgen-receptor levels of both PC-3 cells and DU-145 cells, which are reported to be androgen insensitive [21]. Both cell lines have detectable androgen-receptor levels, which are lower compared with those found in androgen-responsive cells, such as LNCaP [22-24]. Using various concentrations of natural ß-interferon, they found a remarkable enhancement of androgen binding sites, which was dependent on the interferon dose. PC-3 cells exposed to concentrations from 100 to 1,000 IU/ml of the drug showed statistically significant elevations of the androgen binding sites per cell (2-fold and 4-fold, respectively). In the DU-145 cells, receptor enhancement was less pronounced than in PC-3 cells, though significant at the highest interferon concentration (24,261.000 ± 2,177.632 sites per cell compared to 14,793.000 ± 2,076.278 in control cells). In a subsequent study, the same authors investigated the effect of ß-interferon on the androgen-sensitivity of PC-3 cells using 5alpha-dihydrotestosterone for growth stimulation, since testosterone has been shown to be ineffective, probably because of the low level of 5 alpha-reductase present in PC-3 cells [22]. Whereas growth stimulation after 3 and 6 days of exposure to various concentrations of dihydrotestosterone did not occur, significant stimulation was documented in cells pretreated with 1,000 IU/ml ß-interferon both after 3 days and, even more pronounced, after 6 days of exposure to DHT, suggesting that the modification of androgen receptor content might be involved in the modulation of androgen sensitivity. In contrast to these results, Krongrad et al. demonstrated that the mere expression of the human androgen receptor in an androgen unresponsive human prostate cancer cell line (PPC-1) was insufficient to confer androgen responsiveness [24]. They transfected a cDNA clone encoding the human androgen receptor into the PPC-1 line, which lacks detectable expression of androgen receptor or androgen receptor mRNA. Despite the expression of high receptor levels, none of the transfected cell lines demonstrated an alteration of growth rate when incubated with either the androgen mibolerone or the antiandrogen hydroxyflutamide, suggesting that androgen receptor expression is only one component of the pathway that mediates androgen dependent cellular proliferation. Thus, it remains unclear whether receptor modulation obtained in cell cultures is capable of restoring hormone sensitivity *in vivo*.

Antiproliferative Action of Interferons

Cell growth inhibitory action of interferons on various cell lines has been previously established [25-28], and a variety of possible mechanisms involved in growth inhibition have been suggested [29,30]. However, they do not completely elucidate the mechanism(s) by which growth control is achieved. Moreover, it has been suggested that the antiproliferative effect of interferon is reversible, since lowered growth rates were restored to normal levels by removal of interferon, indicating a cytostatic rather than a cytocidal effect [31].

Hayward et al. investigated the effects of all types of interferons on the activities of several enzymes of the carbohydrate metabolism in benign prostatic hyperplasia cells in primary culture [32]. Only the activity of α-glycerolphosphate dehydrogenase (α-GPDH), one of the key enzymes of carbohydrate metabolism involved in the transfer of substrates into the pathways of lipid deposition, thereby regulating energy generation through substrate availability, was modified by interferons. It is understood that an increase in the activity of this particular enzyme indicates deprivation of substrates for energy generation, thus resulting in growth inhibition. They could demonstrate that interferon-α, both in the presence and in the absence of testosterone, increases α-GPDH activity, whereas interferon-γ produced an opposite effect, as did interferon-ß in the absence of testosterone. In the presence of testosterone, interferon-ß exhibited a biphasic effect in that at lowest concentrations the enzyme activity was increased, whereas it was decreased at higher concentrations. This observation gains even more in importance because higher activity of this enzyme has been previously reported to be associated with a lower risk of recurrence in breast carcinoma [33], and the beneficial use of interferon-α in the treatment of prostatic cancer was suggested by the authors.

Provided that substrate availability for energy generation is regulated similarly both in BPH and neoplastic prostatic cells, increased α-GPDH might be responsible for growth inhibition of prostatic cancer cells.

Kimchi proposed that the antiproliferative activity of interferon is exerted by the reduced expression of nuclear oncogene(s) which can be turned on by platelet-derived growth factor (PDGF) [34], which was shown to counteract the suppressive effect of interferon on the expression of c-*myc* [35].

The role of cAMP, which was reported to participate, though depending on the cell type, in the antiproliferative action of interferon [36-38], was investigated by Okutani et al. in the PC-3 cell line [31]. They demonstrated that incubation with interferon-α significantly increased the cellular cAMP level (about 20-fold) and that the cAMP concentrations remained high for at least 24 hours. This elevation paralelled marked growth inhibition. When PGE$_1$, known to increase cAMP concentration by activating adenylate cyclase, was added to the culture, maximum cAMP level was reached within a few minutes, but then decreased rapidly within 2 hours. PGE$_1$, in their experience, did not inhibit PC-3 cell growth, unless added every 24 hours. When dibutyryl cAMP was added to the culture, potent growth inhibition comparable to

that of interferon-α was observed, suggesting that interferon-α exerts its antiproliferative effect on PC-3 cells by continuously elevating the intracellular cAMP level. In addition, the authors could prove that this effect of interferon is not related to the production of TGF-ß by the cells, because TGF-ß concentration in the conditioned medium of PC-3 cells was not affected by interferon, not even at higher doses.

A variety of reports emphasise the growth inhibiting effects of different interferons in both cell cultures and animal models. Comparison of these results needs to be done with caution, since different concentrations of interferons, different endpoints as well as varying culture medium conditions have been employed.

Desphande et al. [39], studying the effects of various interferons on human benign prostatic hyperplasia cells in primary culture, reported that interferon-α inhibited growth of these cells in a dose-dependent manner in that maximal inhibition was seen at the highest doses, irrespective of the presence or absence of testosterone, whereas interferon-ß had little effect at the dose levels used in their study. By contrast, interferon-γ stimulated growth dose dependently, suggesting that this protein acts like other growth factors associated with proliferation. This finding is in disagreement with data reported on malignant cells. Sica et al. investigated the effects of natural ß-interferon and interferon-α on both the DU-145 and the PC-3 cell line [40]. Concentrations used in this experiment ranged from 5 to 1,000 IU/ml. They found that after 3 and 6 days, respectively, both cell lines were inhibited. However, the PC-3 cell line turned out to be more sensitive (60% inhibition by interferon-α, 95% inhibition by interferon-ß) than the DU-145 (16% and 42%, respectively), and natural interferon-ß was found to be more active compared to interferon-α.

Similar to Sica, we found no significant inhibitory effect of interferon-α at various concentrations (1-1,000 IU/ml) in the DU-145 cell line as assessed after 3 days of exposure to the drug. In this experiment we found even increased cell numbers at lower interferon concentrations (Fig. 1). This effect was, however, not statistically significant (unpublished data).

Goldstein et al. investigated the effect of interferon-ß on the growth of DU-145, PC-3 and LNCaP cell lines [41]. Like others, they found growth inhibition in each of the 2 hormone-independent cell lines but no effect in the hormone-

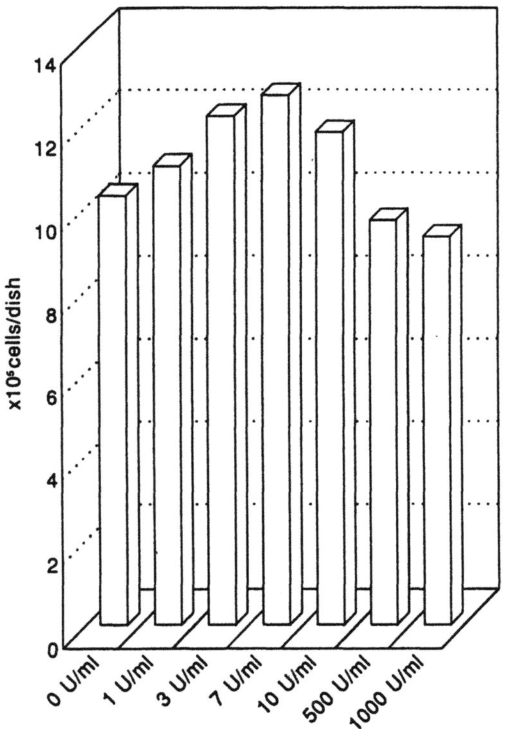

Fig. 1. Effect of various concentrations of interferon-α2b on DU-145 cell line proliferation

dependent cell line LNCaP. However, additional valuable information is provided in this particular paper, concerning both interferon resistance of LNCaP cells and interaction of interferon with growth factors. Interestingly, in LNCaP cells, 2,5A-synthetase, usually an interferon-inducible enzyme, is not induced by interferon-ß. The authors speculated that changes in interferon receptor number or function might be involved in interferon resistance. The antiproliferative effects of interferon-ß in this report could not be reversed by the addition of TGF-α or EGF, and the combination of interferon-ß with TGF-ß1 produced, at best, additive antiproliferative action, although a synergistic inhibitory effect was expected. In contrast to the observed lack of antiproliferative activity of interferon-ß against the LNCaP cell line, van Moorselaar et al. [42] found strong antiproliferative effects of interferon-α against both LNCaP and PC-3 and DU-145 in a nude mouse model, whereas interferon-γ turned out ineffective in the same model. Liu et al., however, demonstrated significant inhibition (80%) of PC-3 proliferation by interferon-γ at a concentration of 500 IU/ml over a 9-day period [43]. However, the combination of interferon and suramin, which has been shown to inhibit the proliferation of prostate cancer in vitro [44], induced less inhibition than interferon alone, suggesting that suramin might interfere with the binding of interferon-γ or otherwise block its cellular receptor. In contrast, Zippe et al. demonstrated that the combination of suramin and interferon-alpha, assessed in both DU-145 and PC-3 cells, was inhibitory in a dose-dependent fashion with an overall increased effect in combination than with either one alone [45]. Fidler et al. examined the antiproliferative properties of 16 recombinant human interferon-α B/D hybrids and found a potent direct antiproliferative activity against a large variety of human tumour lines, emphasising that sensitivity to interferon-α hybrids was independent of sensitivity to interferon-γ [46]. Schmid et al. investigated the antitumour activity of interferon-ß against both PC-3 and DU-145 cells, showing dose-dependent inhibition of growth in both lines [47]. When they combined interferon and estramustine, though suboptimal doses were used, a significant supra-additive inhibition (79%) was seen in the DU-145 cell line compared to 55% (interferon) and 12% (estramustine) inhibition. Accordingly, additive inhibitory effects were found in the PC-3 cells.

Though potentiation of the antitumour activity by the combination of interferon and estramustine was demonstrated, the authors failed to provide a possible explanation for their findings.

In a recent study, van Moorselaar et al. examined the antitumour effects of rat γ-interferon and human tumour necrosis factor alpha (TNF-α) against androgen-dependent and independent Dunning rat prostatic tumours [48]. In their experience, interferon-γ inhibited the PIF-1 cell line at 3 out of 4 concentrations tested, whereas inhibition of the PAT-2 line was found only at the highest concentration. No effect was seen in the MatLyLu line, nor did TNF-α inhibit any of these lines at any concentration. Combining the highest dose of interferon and the highest dose of TNF, significant inhibition occurred in all 3 lines. Substantial augmentation of the efficacy against the androgen-dependent H line in vivo also resulted from combined treatment with the highest dose of interferon-γ and TNF. However, despite a remarkable growth inhibition, tumour regression was not achieved. In contrast to their in vitro findings, interferon-γ monotherapy was very effective against the MatLyLu tumour in vivo and combination with TNF was even more effective. It was also demonstrated that after termination of treatment the tumours regained their initial growth potential after a 10-day lag phase. In terms of survival it turned out that treated animals lived significantly longer than the untreated controls. Nevertheless, the authors concluded from their extensive studies that both cytokines have only limited direct cytotoxic or cytostatic effect on prostatic cancer cell lines. However, TNF exerts its anticellular effect in a similar way as seen in apoptosis, the process of programmed cell death, e.g. following androgen ablation [49,50]. Enhancement of this particular TNF action by interferon-γ has also been reported [51]. In view of this finding, the possible activation of apoptosis in androgen-independent cancer cells by cytokine combination therapy was confirmed.

In another study, van Moorselaar et al. examined the antiproliferative activities of human interferon-α, human interferon-γ and human tumour necrosis factor alone or in combination against 2 prostate cancer xenografts transplanted in nude mice [52]. This model enables to study only the direct antiproliferative effects of cytokines on heterotransplanted human tumours under in vivo conditions. They demonstrated that interferon-α was substantially effective against the PC-3 tumour, but ineffective against the DU-145 tumour. Interferon-γ was less effective than interferon-α, and only at a dose of 80 ng/g significant antitumour activity was recorded. Combination of α and γ-interferon had significant antiproliferative effects against the PC-3 tumour, but not against the DU-145 tumour. Combinations of both interferon-α/TNF and interferon-γ/TNF proved effective in both tumour lines, the former being more effective because complete growth inhibition occurred and no tumours developed after cessation of treatment within a period of at least 8 weeks.

Research during the past few years clearly has demonstrated that interferons do exert some antiproliferative effect in prostatic cancer cells, though the basic mechanism has not yet been elucidated. To explain the, at least partly, conflicting data published so far, it might be' assumed that the susceptibility of cells to interferon varies within the cell cycle. Thus, when non-synchronous cells are used, heterogeneous effects to identical stimuli might result. In addition, as a possible cause of artefacts, the composition of fetal calf sera added to most cultures is likely to differ with respect to steroid, growth factor or cytokine content, thus suggesting that serum-free culture medium should be used for future investigations.

Immunomodulatory Effects of Interferon

Interferon therapy in solid tumours, particularly in prostatic carcinoma, may not be as successful as expected, because most of the tumours are weakly antigenic and they do not express the strong antigens that can produce vigorous immune response capable of rapidly destroying a tumour.

Evidence for the relevance of the immune system in prostatic cancer is given by Mickey et al. who demonstrated in an animal model that immunomodulators can diminish tumour-induced immune suppression and lead to retarded tumour growth, decreased metastatic spread and, consequently, prolonged survival [53]. Further support for the assumed significance of the immune system comes from Romijn et al., who investigated the survival of PC-3 cells in an ani-

mal model and concluded that immune defense mechanisms are at least partly responsible for tumour-cell rejection [54].

The role of natural killer (NK) cell activity has been widely investigated in prostatic cancer. There is consensus in the literature that in patients with advanced disease the level of NK activity is reduced, which is not associated with the age of the patients. Patients with localised disease were shown to have either increased [55] or normal activity [56,57] as compared to healthy controls. In addition, differences between the activity in peripheral blood cells and periprostatic lymph-node cells have been reported [58].

Choe et al. [57] demonstrated that NK activity of advanced prostatic cancer patients significantly improved after *in vitro* incubation with interferon-α, suggesting that the lower NK activity is not due to a depletion of the NK cell precursor population. Although the basic mechanism of this interferon-induced augmentation of NK activity is not completely understood, it might be speculated that *in vivo* interferon therapy can be beneficial by activating NK precursor cells to express their cytotoxic potential. In a similar study, Schwemmer et al. [55] demonstrated that NK activity can be enhanced by interferon-ß both in healthy controls and in patients with different stages of cancer, though enhancement in patients with advanced disease is less pronounced. They also demonstrated that within the group of patients with advanced disease those with progression had barely detectable NK activity and their ability to respond to interferon stimulation was more reduced than in those with stable disease, suggesting that tumour burden and NK cell-mediated cytotoxicity are inversely related. However, it remains unclear whether NK cell activity fails prior to metastatic spread or is inhibited because of metastatic disease. Wirth et al. [58] also demonstrated that NK cell activity can be restored by interferon-ß in patients with advanced disease. They consequently concluded that this is indicative of a normal target structure recognition on the tumour cell membrane.

In a model system of murine metastases, it has been demonstrated that the number of metastases can be decreased if interferons are given prior to intravenous tumour cell challenge, suggesting an increased NK cell activity [59]. Studies in murine cell lines resistant to the growth inhibitory effects of interferons *in vitro* evidenced that interferons do exert antitumour activities in such cell lines, even when they lack interferon receptors [60]. These observations suggest that host immune effector cells are stimulated by interferons, hence leading to elimination of the tumour, even though demonstration of NK cell activity against primary tumours has failed so far.

In contrast, the role of NK cells in inhibiting the metastatic process is critically discussed by Wade et al. [61]. Using the PA-11 transplantable prostate adenocarcinoma in the Lobund-Wistar rat, she demonstrated that NK cells, in this particular model, have no inhibitory effect on metastases.

Further evidence for the involvement of an impaired immune system in cancer was recently provided by Mizoguchi et al. [62], who demonstrated that animals with colon cancer develop CD8+ T cells with impaired cytotoxic function when they bear a tumour longer than 26 days. Whether this is true also for prostatic cancer needs to be investigated.

Currently the importance of cell surface antigens, such as major histocompatibility complex (MHC) antigens, for the induction of T cell response has been recognised. In this context, it is important to know that helper T cells are MHC class II antigen restricted, and cytotoxic T cells are MHC class I restricted. It is therefore tempting to speculate that cytokines, especially interferon-γ, may have an antitumour effect because they upregulate class I and class II antigens in some tumour cells. However, knowledge about their significance in prostate cancer is very limited. It has been shown that normal prostatic glands express class I but not class II antigen [63]. In a recent study, Bigotti et al. [64] investigated immunohistochemically the specimens of 38 patients with prostatic adenocarcinoma. They demonstrated that HLA class II-positive carcinomas were low grade and associated with a more favourable prognosis than those staining negatively. A similar result for papillary carcinoma of the thyroid has been previously published by Goldsmith et al. [65]. To make things even more confusing, a worse prognosis associated with positive staining for HLA class II antigens has been shown in melanomas [66]. It is therefore deemed appropriate to further study the significance of these antigens in larger series.

In this context, we studied the DU-145 cell line with respect to the cells' genetic information concerning HLA class II antigens (unpublished

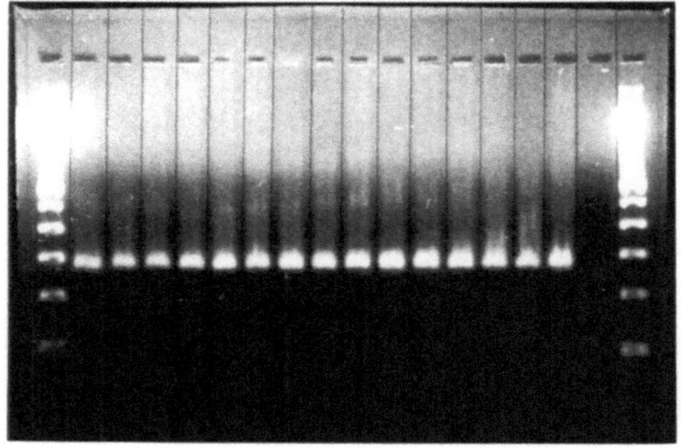

Fig. 2. Submarine agarose gel electrophoresis

data). Employing the PCR SSO method (polymerase chain reaction, sequence-specific oligonucleotide probes) [67], we could demonstrate that the information DR 1 is present within the genome of DU-145 cells. The PCR products are visualised by submarine agarose gel electrophoresis (Fig. 2) and further analysed by testing hybridisation activity with a series of synthetic sequence-specific oligonucleotide probes. In the case of an absolute nucleotide sequence match between the SSO probe and the membrane-bound DNA, a positive signal is shown on a luminescence detection film (Fig. 3). Interaction of the T cell with an antigen-presenting cell may be further enhanced by the expression of adhesion molecules such as LFA-3 (leukocyte function associated antigen) and ICAM-1 (intercellular adhesion molecule). Graham et al. [68] studied the differential expression of ICAM-1 in patients with benign and malignant prostatic disease. Whereas all specimens from BPH tissue demonstrated cell surface staining of ICAM-1 in prostatic acinar cells, 5/6 neoplastic tissues failed to reveal such staining. The one carcinoma with ICAM-1 expression was well differentiated. They concluded that the loss of expression of ICAM-1 might be an important factor in the loss of immune surveillance or the development of uncontrolled growth.

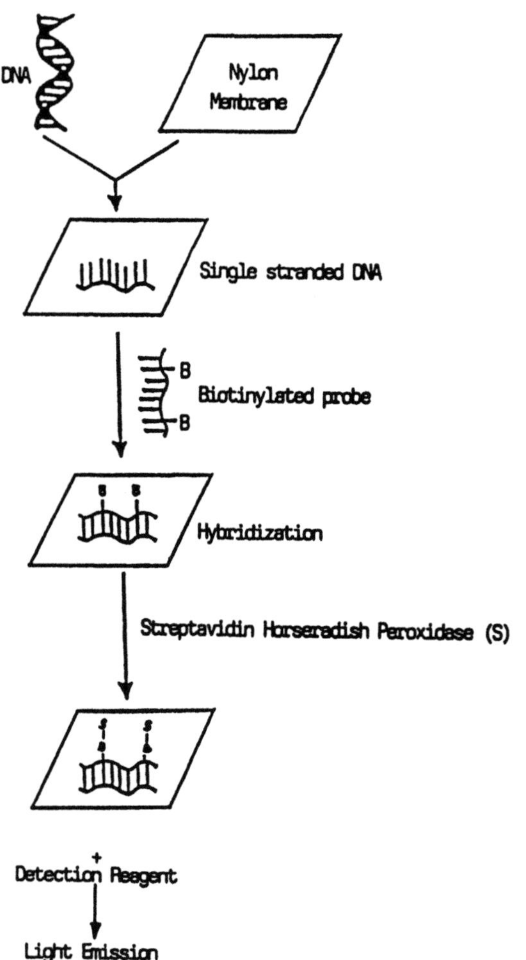

Fig. 3. Non-radioactive detection with biotinylated probes

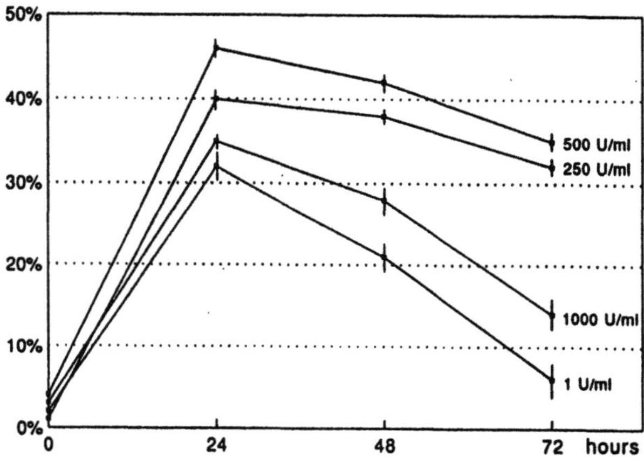

Fig. 4. ICAM-1 expression in DU-145 cells after 24, 48 and 72 hours of treatment with various concentrations of interferon-α 2b

In our own experience (unpublished data), we could demonstrate that ICAM-1 expression in the DU-145 cell line is enhanced at various concentrations of interferon-α (Fig. 4). However, expression is less pronounced both at the highest and the lowest concentration used, suggesting that higher doses do not necessarily imply maximal effectiveness.

In general, immunomodulatory effects of interferons in prostatic cancer cells have not been extensively investigated up to now and their clinical significance remains to be established. However, the development of more sophisticated techniques is promising for future research work.

Interferons in Clinical Trials

Since their discovery in 1957, interferons underwent extensive clinical testing. Early in the 1980s, when recombinant preparations became available, the assessment of their therapeutic efficacy in a wide range of malignancies began with many thousands of cancer patients being treated in phase I, II, and III trials. While the impact of interferons in haematopoietic and lymphatic malignancies is already well established, their role in solid tumours is not well elucidated yet. This is particularly true for prostate cancer, which was, so far, only superficially explored as to its susceptibility to interferons (Table 1). During the past 10 years, only a limited number of patients with prostate cancer, in most instances suffering from far advanced disease and with multiple pretreatment, were subject to interferon administration, either as a single agent or in combination with hormonal and/or chemotherapy. The clinical experience of the few published series, some of them representing only anecdotal observations, will be reviewed in this section. In general, the results are disappointing, responses were reported infrequently, and toxicity was substantial in some studies. However, no firm conclusions can be drawn at present because the number of patients entered in each study has been too small, different types and preparations of interferons have been employed, different doses with dose adjustment in many cases have been used, schedules, route of administration, as well as duration of treatment have been different,

Table 1. Clinical trials with interferon in prostate cancer

Interferon	Phase	Pts/eval.	Response	Ref.
lym alpha	I	2/2		[70]
α_{2b} + doxo	I	1/0		[71]
α_{2b} + 5FU	1	1/1	1 SD	[78]
α_{2b}	II	9/9	1 mixed response	[72]
α_{2b}	II	25/20	1 CR, 1 PR	[73]
α_{2b}	II	8/8	no response	[77]
Le-IFN	II	10/10	1 CR, 3 PR	[74]
Le-IFN	II	1/1	1 PR	[75]
Le-IFN	II	14/13	5 CR, 6PR	[76]
α_{2b} + 5FU	II	13/13	3 subj. responses	[79]
ß	II	16/16	3 SD	[81]
γ	I	?		[82]

and the response criteria have rarely been defined.

Most of the trials employed interferon-α, the effectiveness of which has been suggested in a large number of preclinical studies, as shown in the previous sections. There is some evidence that interferon-α, in man, exerts its antitumour activity by growth regulation rather than by immunostimulation [69]. Certainly, interferon-α therapy can enhance NK cell cytotoxicity in peripheral blood, but such changes are not related to response to therapy.

The first phase I study, in which 2 patients with prostate cancer were included, was published by Laszlo et al. in 1983 [70]. In this extensive biphasic study, the pharmacological and immunological effects of human lymphoblastoid interferon in cancer patients were evaluated and important information concerning side effects after both single-dose (phase A of the study) and chronic (phase B) drug administration is provided. The major toxicities, fever, chills, fatigue, anorexia, malaise, myalgia, and weight loss, were encountered during chronic drug administration. Dose-limiting toxicity was 15×10^6 IU/m^2 for repeated intramuscular injections, but some patients with a more rapid dose escalation scheme tolerated up to 35×10^6 IU/m^2. The authors emphasise that virtually all patients managed to reach approximately the same total dose of 135×10^6 units of interferon, when it was given in the chosen schedule.

A more recent phase I study investigating the combination of recombinant human interferon-α_{2a} with escalating doses of doxorubicin also included one patient with prostatic cancer [71]. This patient, 70 years old with ECOG performance score 0, experienced severe, diffuse osseous pain within one day of the initial interferon injection (12×10^6 IU/m^2 i.m.) and had to be removed from the study. Unfortunately, there is no further information given whether or not this patient developed progressive disease along with bone pain.

Chang et al. prematurely closed a phase II study in patients with hormone-resistant prostate carcinoma due to intolerable grade III and IV toxicities resulting in deterioration of performance status [72]. They treated 9 patients with either 5×10^6 IU/m^2 (patients older than 60 years) or 10×10^6 IU/m^2 human recombinant interferon-α_{2b} subcutaneously 3 times a week. The median total dose in this series was 153×10^6 IU, which is consistent with the data given by Laszlo. Only one mixed response was recorded and the treatment was thus deemed inappropriate. A possible explanation for the unusually high toxicity in this study (7 patients exhibited neurotoxicity, 7 patients became bedridden or required hospitalisation) might be the fact that, in general, the performance status of these patients at baseline was poor, the patients exhibited extensive bone metastases, and 6/9 patients had undergone more than one previous treatments.

Another phase II study enrolling 25 patients with metastatic hormone-resistant prostate cancer is reported by van Haelst-Pisani et al. [73]. Using a similar dose schedule of 10×10^6 U/m^2 3 times a week, toxicity, though less pronounced, was also substantial and dose reduction was required in all but 3 patients. Out of 20 evaluable patients 13 had progressed after 2 months of treatment and 4 patients had stable disease for 5 to 7 months. One patient had a partial response with his PSA dropping from 6,400 to 212 ng/ml and signs of retroperitoneal lymph-node regression, but this patient refused further treatment. One complete remission was obtained for 9+ months. In this patient PSA fell to an undetectable level and complete regression of retroperitoneal lymph nodes occurred.

Another group of 10 patients with metastatic disease was treated with partially pure léukocyte interferon by Gutterman and Quesada [74]. They observed subjective or objective improvement in 4 patients, one of them accomplished complete remission. Unfortunately, no information is provided concerning the baseline status of the patients, dose of interferon, treatment duration or time to response. However, the authors generally stated that prostate cancer along with colon, ovarian, and kidney cancer appeared, at least in their experience, less sensitive than other tumours.

Peddinani and Savery treated one patient with metastatic progressive prostatic cancer with human leukocyte interferon at a dose of 7 to 9 x 10^6 IU daily for 12 weeks [75]. His acid phosphatase returned to normal and a relief of bone pain was achieved. Despite the comparingly high dose, no treatment-related side effects were recorded with the exception of one brief episode of mild fever. Duration of response is not provided in this report.

The most spectacular series with respect to leukocyte interferon-induced regression of metastatic cancer resistant to chemo- and/or radiotherapy is published by Medenica and Slack [76]. Among other tumours they treated 14 patients (mean age: 53) with disseminated prostatic cancer. Complete remission was achieved in 5 patients (mean duration: 13 months), partial remission in 6 patients (mean duration: 10 months). The overall objective response rate (in 13 evaluable patients) was 84%, one additional patient remained stable for 10 months. This study differs from others in many aspects. The most striking feature is that a pulse therapy schedule was employed. Patients received 6×10^6 IU/m^2 interferon intramuscularly for 3 consecutive days every 4 weeks. Another important point is that the duration of treatment exceeded one year, which is consistent with preclinical observations [31,48] suggesting prolonged treatment periods. However, the observed time to treatment response was surprisingly short in this series with an average time of 86 days (CR) and 56 days (PR), respectively. Toxicity was tolerable and mostly comprised the typical flu-like symptoms in the majority of the patients. There was a clear correlation between fever and response in that higher fever was predictive for objective clinical response and the advisability of antipyretic drugs along with interferon therapy was questioned. Though in solid tumours the antiproliferative rather than the immunomodulatory mode of action of interferon is presumed, one might speculate that the outstanding results depicted in this study are due to a greater enhancement of the NK activity by the pulse schedule, particularly since it has been shown that chronic administration of higher doses of interferon depresses NK activity [70].

In a series of 8 patients with multiple pretreated progressive disease, Goldschmidt et al. [77] found no objective response to 5×10^6 IU interferon-α 3 times a week. In a recent phase I study investigating the combination of recombinant human interferon-α_{2b} (5-10 x 10^6 IU s.c. 3 times a week for 6 weeks), 5-fluorouracil and leucovorin, one 80-year old patient with hormone-resistant prostate cancer was enrolled [78]. This patient showed a 50% decrease in PSA and his measurable pelvic mass remained stable by CT scan, while toxicity was tolerable. A similar regimen with interferon-α_{2b} (9 x 10^6 IU s.c. 3 times a week for 8 weeks) and 5-fluorouracil was used by Dreicer et al. [79] in 13 patients with hormone-refractory metastatic prostate cancer. In contrast to Taylor's observation, toxicity was moderate to severe, and no objective response was seen. Only one patient's bone scan improved and 3 patients noted improvement in bone pain. It is of interest that at the beginning of treatment most patients (82%) had a transient rise in PSA up to 77% above baseline, some of them in discordance with clinical improvement.

In an ongoing study (unpublished data), we treated 6 symptomatic patients with advanced prostatic cancer who had failed on prior hormonal therapy, with interferon-α_{2b} 3 x 10^6 IU s.c. 3 times weekly in combination with either flutamide, cyproterone acetate or estramustine phosphate. In this group, 2 patients subjectively responded with an impressive relief of bone pain. One of them also showed an objective response with an improvement of bone scan and a significant drop of PSA from 309 to 14.8 ng/ml, the other patient exhibited stabilisation for 11+ months. Four patients progressed during therapy. Toxicity was moderate and typically comprised fever and flu-like symptoms. In these patients PSA, TPA, as well as the soluble adhesins E-selectin and ICAM-1 were measured, the latter two having been considered to give evidence of endothelial damage [80] as expected in progressing tumours. Both E-selectin and ICAM-1 decreased in responding patients (Table 2), thus suggesting tumour-inactivation by interferon therapy.

Though preclinical studies evidenced that interferon-ß is at least as effective as interferon-α with respect to its growth inhibitory capacity in prostatic cancer cell lines, there exists only one clinical trial of interferon-ß in the treatment of advanced, hormone-refractory prostate cancer [81]. This study conducted by the National Prostate Cancer Project (NPCP Protocol 2100) enrolled 16 patients and was prematurely closed, because no objective response was noted and only limited efficacy was concluded. Patients received 6 x 10^6 IU interferon-ß intravenously 3 times weekly for 12 weeks. While toxicity was considered moderate, 13 patients progressed, 8 of whom during therapy, and 3 patients had stabilisation for a mean of 6 months.

No clinical trial employing interferon-γ in the treatment of prostate cancer has been published so far. There is only a single phase I trial [82] in which prostate cancer patients have been entered. Treatment comprised recombinant interferon-γ with doses increased stepwise from 0.5 to 10 x 10^6 JRU (Japan Reference Unit) for 5 consecutive days every 3 weeks. All side effects such as fever, chills, nausea, vomiting, fatigue, bone marrow suppression, and hypotension were transient and not life-threatening, but occurred at each step. No data concerning efficacy are given.

Table 2. Influence of interferon therapy on prostate specific antigen (PSA), tissue polypeptide antigen (TPA) and the soluble adhesins E-selectin and ICAM-1

	PSA	TPA	E-selctin	ICAM-1
responder	↓	↓	↓	↓
non-responder	↑	↑	↑	↓

It is impossible, at present, to decide whether there is a place for treating prostate cancer with interferon and the studies reviewed above raise more questions than they are able to answer.

The differences in patients' characteristics and treatment schedules may account for the divergent results both with respect to efficacy and toxicity. Further clinical investigations need to clearly define the target population for cytokine therapy. As yet it is not clear whether previously untreated or pretreated patients are candidates for interferon therapy, nor has the ideal clinical stage been defined. In this context, it would be worthwhile to investigate patients with clinical stage C prostatic cancer, in which cure either by radical surgery, radiation therapy, or conventional hormonal treatment is unlikely, but general constitutional conditions would allow prolonged treatment periods. This is even more true, since the potential use of interferons in the control and elimination of metastatic disease has been shown at least in animal models [59,83, 84].

Patients' selection within a certain clinical stage of disease might also be of importance. Horn et al. [85] measured serum interferon levels as a marker in patients with malignant tumours. They demonstrated that serum levels were higher in patients with active disease (as defined by clinical examination, laboratory tests, or radiological studies) than in those with inactive disease. In prostate cancer patients they found significantly elevated levels at a late stage. Furthermore, Kita et al. [86] measured the interferon-α producing capacity in 47 patients with various stages of prostate cancer. They found a tendency to decrease as the stage of the tumour was more advanced and could correlate a low producing capacity with a poorer prognosis.

In addition, no clear-cut guidelines concerning the dose, the dose schedule, route of administration, and even time of administration do actually exist. As far as the latter is concerned, the observation might be important that with equal doses of interferon given in the morning, the serum level obtained was lower than it was when interferon was given in the afternoon to the same patient [76].

Finally, there is a tremendous need to establish approaches in combination either with hormonal, cytostatic, or biological measures to encounter the sequential developmental process of cancer. However, in expectation of future progress in the field of fighting prostate cancer by means of interferon or other cytokine therapy, one should be alert to a paper recently published by Erlacher et al. [87]. They successfully treated a patient with hairy cell leukaemia by long-term administration of interferon and complete remission was maintained for 6 years. However, 2 years after the start of treatment, a metastasising carcinoma of the prostate was diagnosed, leading to death after 4 years.

REFERENCES

1   Boring CC, Squires TS, Tong T: Cancer statistics. Ca-A Cancer J Phys 1992 (42):19-38

2   Edwards CW, Steinthorsson E, Nicholson D: An autopsy study of latent prostatic cancer. Cancer 1953 (6):531-554

3   Scardino PT: Early detection of prostate cancer. Urol Clin N Am 1989 (16):635-655

4   Hanash KA: Natural history of prostatic cancer. In: Murphy GP, Khoury S, Küss R, Chatelain C, Denis L (eds) Progress in Clinical and Biological Research. Alan R Liss Inc, New York 1987 (Vol 243 B) pp 289-320

5   Veterans Administration Cooperative Urological Research Group (VACURG): Treatment and survival of patients with cancer of the prostate. Surg Gynecol Obstet 1967 (124):1011-1017

6   Huggins C, Hodges CV: Studies of prostatic cancer: I. Effect of castration, estrogen and androgen injections on serum phosphatase in metastatic carcinoma of the prostate. Cancer Res 1941 (1):293-297

7   Labrie F, DuPont A, Belanger A: Complete androgen blockade for the treatment of prostate cancer. In: DeVita VT Jr, Hellman S, Rosenberg SA (eds) Important Advances in Oncology. JB Lippincott, Philadelphia 1985 pp 193-217

8   Crawford ED, Eisenberger MA, McLeod DG, Spaulding JT, Benson R, Dorr FA, Blumenstein BA, Davis MA, Goodman PJ: A controlled trial of leuprolide with and without flutamide in prostatic carcinoma. N Engl J Med 1989 (321):419 -424

9   Isaacs JT: The timing of androgen ablation therapy and/or chemotherapy in the treatment of prostatic cancer. The Prostate 1984 (5):1-17

10 Pummer K: Epirubicin plus flutamide and orchidectomy in previously untreated advanced prostatic cancer. Seminars in Oncology 1991 (18, suppl 6):26-28

11 Balkwill FR: Interferons. In: Cytokines in Cancer Therapy. Oxford University Press, Oxford 1989 pp 8-53

12 Gonor SE, Lakey WH, McBlain WA: Relationship between concentrations of extractable and matrix-bound nuclear androgen receptor and clinical response to endocrine therapy for prostatic adenocarcinoma. J Urol 1984 (131):1196-1201

13 Katzenellenbogen BS, Kendra KL, Norman MJ, Berthois Y: Proliferation, hormonal responsiveness, and estrogen receptor content of MCF-7 human breast cancer cells grown in the short-term and long-term absence of estrogens. Cancer Res 1987 (47):4355-4360

14 King RJB: Receptors, growth factors and steroid insensitivity of tumours. J Endocrinol 1990 (124):179-181

15 Trapman J, Brinkmann AO: Properties and expression of the androgen receptor in prostate cancer. In: Berns PMJJ, Romijn JC, Schröder FH (eds) Mechanism of Progression to Hormone-Independent Growth of Breast and Prostatic Cancer. Casterton Hall, Carnforth, The Parthenon Publishing Group Limited, Lancs 1991 pp 153-168

16 Ekman P, Brolin J: Steroid receptor profile in human prostate cancer metastases as compared with primary prostatic carcinoma. The Prostate 1991 (18):147-153

17 van den Berg HW, Leahey WJ, Lynch M, Clarke R, Nelson J: Recombinant human interferon alpha increases oestrogen receptor expression in human breast cancer cells (ZR-75-1) and sensitizes them to the antiproliferative effects of tamoxifen. Br J Cancer 1987 (55):255-257

18 Sica G, Natoli V, Stella C, Del Bianco S: Effect of natural beta-interferon on cell proliferation and steroid receptor level in human breast cancer cells. Cancer 1987 (60):2419-2423

19 Kangas L, Nieminen AL, Cantell K: Additive and synergistic effects of a novel antiestrogen, toremifene (Fc-1157a), and human interferons on estrogen responsive MCF-7 cells in vitro. Med Biol 1985 (63):187-190

20 Goldstein D, Bushmeyer SM, Witt PL, Jordan VC, Borden EC: Effects of type I and II interferons on cultured human breast cells: Interaction with estrogen receptors and with tamoxifen. Cancer Res 1989 (49):2698-2702

21 Sica G, Fabbroni L, Dell'Acqua G, Iacopino F, Marchetti P, Cacciatore M, Pavone-Macaluso M: Natural ß-Interferon and androgen receptors in prostatic cancer cells. Urol Int 1991 (46):159-162

22 Kaighn ME, Shankar Narayan K, Ohnuki Y, Lechner JF, Jones LW: Establishment and characterization of a human prostatic carcinoma cell line (PC-3). Invest Urol 1979 (17):16-23

23 Stone KR, Mickey DD, Wunderli H, Mickey GH, Paulson DF: Isolation of a human prostate carcinoma cell line (DU-145). Int J Cancer 1978 (21):274-281

24 Krongrad A, Allman DR, Brothman AR, McPhaul MJ: The expression of the human androgen receptor (AR) in an androgen unresponsive human prostate cancer cell line (PPC-1) is insufficient to confer androgen responsiveness. J Urol (Abstr LXXXVIth AUA Annual Meeting, Toronto 1991) 1991 (145):295A

25 Borden EC, Balkwill FR: Preclinical and clinical studies of interferons and interferon inducers in breast cancer. Cancer 1984 (53):783-788

26 Borden EC, Groveman DS, Nasu T, Dernikoff C, Bryan GT: Antiproliferative activities of interferons against human bladder carcinoma cell lines in vitro. J Urol 1984 (132):800-803

27 Creasey AA, Bartholomew JL, Merigan TL: Role of G0-G1 arrest in the inhibition of tumour cell growth by interferon. Proc Natl Acad Sci USA 1980 :1471-1475

28 Epstein L, Shen JT, Abele JS, Reese CC: Sensitivity of human ovarian carcinoma cells to interferon and other antitumour agents as assessed by an in vitro semi-solid agar technique. Ann NY Acad Sci 1980 (350):228-244

29 Fuse A, Kuwata T: Inhibition of DNA synthesis and alteration of cyclic adenosine 3',5' monophosphate levels in RSa cells by human leukocyte interferon. J Natl Cancer Inst 1978 (60):1227-1232

30 Tovey M, Rocheete-Egly C, Castagna M: Effect of interferon on concentrations of cyclic nucleotides in cultured cells. Proc Natl Acad Sci USA 1979 (76):3890-3893

31 Okutani T, Nishi N, Kagawa Y, Takasuga H, Takenaka I, Usui T, Wada F: Role of cyclic AMP and polypeptide growth regulators in growth inhibition by interferon in PC-3 cells. The Prostate 1991 (18):73-80

32 Hayward S, Cox S, Mitchell I, Hallowes R, Desphande N, Towler J: The effects of interferons on the activity of glycerolphosphate dehydrogenase in benign prostatic hyperplasia cells in primary culture. J Urol 1987 (138):648-653

33 Desphande N, Mitchell I, Millis R: Tumour enzymes and prognosis in human breast cancer. Eur J Cancer 1981 (17):493-496

34 Kimchi A: Reduction of nuclear oncogene expression by endogenous and exogenous interferons. In: Smyth JD (ed) Interferons in Oncology. Springer Verlag, Berlin 1987 pp 53-65

35 Einat M, Resnitzky D, Kimchi A: Inhibitory effects of interferon on the expression of genes regulated by platelet-derived growth factor. Proc Natl Acad Sci USA 1985 (82):7608-7612

36 Tovey MG: Interferon and cyclic nucleotides. Interferon 1982 (4):23-46

37 Schneck J, Ragen-Zisman B, Rosen OM, Bloom BR: Genetic analysis of the role of cAMP in mediating effect of interferon. Proc Natl Acad Sci USA 1982 (79):1879-1883

38 Weetman AP: Recombinant γ-interferon stimulates iodide uptake and cyclic AMP production by the FTRL5 thyroid cell line. FEBS Lett 1987 (221):91-94

39 Desphande N, Hallowes RC, Cox S, Mitchell I, Hayward S, Towler JM: Divergent effects of interferons on the growth of human benign prostatic hyperplasia cells in primary culture. J Urol 1989 (141):157-160

40 Sica G, Fabbroni L, Castagnetta L, Cacciatore M, Pavone-Macaluso M: Antiproliferative effect of interferons on human prostate carcinoma cell lines. Urol Res 1989 (17):111-115

41 Goldstein D, O'Leary M, Mitchen J, Borden EC, Wilding G: Effects of interferon-ßser and transforming growth factor ß on prostatic cell lines. J Urol 1991 (146):1173-1177

42 Van Moorselaar RJA, Beniers AJMC, Hendriks BT, Van Stratum P, Van der Meide PH, Debruyne FMJ, Schalken JA: In vivo antiproliferative effects of alpha- and gamma-interferon and tumor necrosis factor alpha on androgen-dependent and -independent prostatic tumors. J Urol (Abstr LXXXVth AUA Annual Meeting, New Orleans 1990) 1990 (143):243A

43 Liu S, Ewing MW, Anglard P, Trahan E, La Rocca RV, Myers CE, Linehan WM: The effect of suramin, tumor necrosis factor and interferon-γ on human prostate carcinoma. J Urol 1991 (145):389-392

44 La Rocca RV, Danesi R, Cooper MR, Jamis-Dow C, Ewing MW, Liu S, Linehan WM, Myers CE: Effect of suramin on human prostate cancer cells in vitro. J Urol 1991 (145):393-398

45 Zippe CD, Henry ET, Heston WDW, Fair WR: Antiproliferative effects of suramin, tumor necrosis factor and interferon-alpha on human prostatic carcinoma cell lines. J Urol (Abstr LXXXVth AUA Annual Meeting, New Orleans 1990) 1990 (143):241A

46 Fidler IJ, Heicappell R, Saiki I, Grutter MG, Horisberger MA, Nuesch J: Direct antiproliferative effects of recombinant human interferon-alpha B/D hybrids on human tumor cell lines. Cancer Res 1987 (47):2020-2027

47 Schmid SM, Harrison SD, Cummings KB: Augmented anticellular effects of estramustine combined with interferon beta serine treatment for prostatic and renal carcinoma cells. J Urol (Abstr LXXXIVth AUA Annual Meeting, Dallas 1989) 1989 (141):518

48 Van Moorselaar RJA, Hendriks BT, Van Stratum P, Van der Meide P, Debruyne FMJ, Schalken JA: Synergistic antitumor effects of rat γ-interferon and human tumor necrosis factor alpha against androgen-dependent and -independent rat prostatic tumors. Cancer Res 1991 (51):2329-2334

49 Kyprianou N, English HF, Isaacs JT: Programmed cell death during regression of PC-82 human prostate cancer following androgen ablation. Cancer Res 1990 (50):3748-3753

50 Rubin BY, Smith LJ, Hellermann GR, Lunn RM, Richardson NK, Anderson SL: Correlation between the anticellular and DNA fragmentation activities of tumor necrosis factor. Cancer Res 1988 (48):6006-6010

51 Dealtry GB, Naylor MS, Fiers W, Balkwill FR: DNA fragmentation and cytotoxicity caused by tumor necrosis factor is enhanced by interferon-γ. Eur J Immunol 1987 (17):689-693

52 Van Moorselaar RJA, Van Stratum P, Borm G, Debruyne FMJ, Schalken JA: Differential antiproliferative activities of alpha- and gamma-interferon and tumor necrosis factor alone or in combinations against two prostate cancer xenografts transplanted in nude mice. The Prostate 1991 (18):331-344

53 Mickey DD, Carvalho L, Foulkes K: Combined therapeutic effects of conventional agents and an immunomodulator, PSK, on rat prostatic adenocarcinoma. J Urol 1989 (142):1594-1598

54 Romijn JC, Verkoelen CF, Schroeder FH: Measurement of the survival of human tumor cells after implantation in athymic nude mice. Int J Cancer 1986 (38):97-101

55 Schwemmer B, Lehmer A, Hofmann R, Braun J: Natural killer cell activity in patients with prostatic carcinoma and its in vivo boosting with Bacillus Calmette-Guerin. Urol Int 1984 (39):321-326

56 Schmitz-Dräger BJ, Marumo K, Ackermann R: Kinetik der spontanen zellvermittelten Zytotoxizität bei Patienten mit Prostatakarzinom. In: Harzmann R, Jacobi GH, Weissbach L (eds) Experimentelle Urologie. Springer Verlag, Berlin-Heidelberg 1985 pp 494-500

57 Choe BK, Frost P, Morrison MK, Rose NR: Natural killer cell activity of prostatic cancer patients. Cancer Investigation 1987 (5):285-291

58 Wirth M, Schmitz-Dräger BJ, Ackermann R: Functional properties of natural killer cells in carcinoma of the prostate. J Urol 1985 (133):973-978

59 Hanna N, Fidler IJ: Role of natural killer cells in the destruction of circulating tumor emboli. J Natl Cancer Inst 1980 (65):801-809

60 Gresser I: How does interferon inhibit tumor growth? Interferon 1985 (6):93-126

61 Wade AC, Schmidt MA, Pollard M: The relationship of natural killer cells to metastasis of transplantable prostate adenocarcinoma. The Prostate 1985 (7):53-61

62 Mizoguchi H, O'Shea JJ, Longo DL, Loeffler CM, McVicar DW, Ochoa AC: Alterations in signal transduction molecules in T lymphocytes from tumor-bearing mice. Science 1992 (258):1795-1798

63 Daar AS, Fuggle SV, Fabre JW, Ting A, Morris PJ: The detailed distribution of MHC class II antigens in normal human organs. Transplant 1984 (38):293-298

64 Bigotti G, Coli A, Castagnola D: Distribution of Langerhans cells and HLA class II molecules in prostatic carcinomas of different histopathological grade. The Prostate 1991 (19):73-87

65 Goldsmith NK, Dikman S, Bermas B, Davies TF, Roman SH: HLA class II antigen expression and the autoimmune thyroid response in patients with benign and malignant thyroid tumors. Clin Immunol Immunopathol 1988 (48):161-173

66 Natali PG, Bigotti A, Cavaliere R, Ruiter DJ, Ferrone S: HLA class II antigens synthesized by melanoma cells. Cancer Rev 1987 (9):1-34

67 Mullis KB and Faloona F: Specific synthesis of DNA in vitro via a polymerase-catalysed chain reaction. Methods Enzymol 1987 (155):335-350

68 Graham Jr SD, Metz R, Gal A, Manos S, Swerlick RA: Differential expression of ICAM-1 expression in BPH and prostatic adenocarcinoma. J Urol (LXXXVIth AUA Annual Meeting, Toronto 1991) 1991 (145):350A

69 Smyth JF: Guidelines for the use of INTRON A (interferon alpha-2B) in clinical oncology: lessions from the laboratory. Cancer Treat Rev 1988 (15):3-6

70 Laszlo J, Huang AT, Brenckman WD, Jeffs C, Koren H, Cianciolo G, Metzgar R, Cashdollar W, Cox E, Buckley III CE, Tso CY, Lucas Jr VS: Phase I study of pharmacological and immunological effects of human lymphoblastoid interferon given to patients with cancer. Cancer Res 1983 (43):4458-4466

71 Creagan ET, Frytak S, Long HJ, Kvols LK: Phase I study of recombinant leukocyte A interferon (IFN-alpha2A, Roferon-A) with doxorubicin in advanced malignant disease. Cancer 1989 (64):1034-1037

72 Chang AYC, Fisher HAG, Spiers ASD, Boros L: Toxicities of human recombinant interferon-alpha2 in patients with advanced prostate carcinoma. J Interferon Res 1986 (6):713-715

73 Van Haelst-Pisani C, Richardson RL, Thorneau TM, Hahn RG, Burch PA, Buckner JC, Frytak S: Phase II study of recombinant leukocyte A human interferon Roferon-A (IFL-RA) in patients with advanced hormone-resistant prostate cancer. Am Soc Clin Oncol (ASCO), Washington, DC 1990

74 Gutterman J, Quesada J: Clinical investigation of partially pure and recombinant DNA derived leukocyte interferon in human cancer. Tex Rep Biol Med 1981-1982 (41):626-633

75 Peddinani MV, Savery F: Human leukocyte interferon in the treatment of selected cancer patients. Clin Ther 1983 (5):336-341

76 Medenica R, Slack N: Clinical results of leukocyte interferon-induced tumor regression in resistant human metastatic cancer resistant to chemotherapy and/or radiotherapy - pulse therapy schedule. Cancer Drug Deliv 1985 (2):53-76

77 Goldschmidt AJW, Bauer L, Tunn UW: "Third-line"-Therapie des metastasierten Prostatakarzinoms mit Interferon. In: Hofstetter A, Staehler G, Kriegmair M, Schuth J (eds) Zytokine in der Urologischen Onkologie. W. Zuckschwerdt Verlag, München 1990 pp 73-82

78 Taylor CW, Modiano MR, Woodson ME, Marcus SG, Alberts DS, Hersh EM: A phase I trial of fluorouracil, leucovorin, and recombinant interferon alpha-2b in patients with advanced malignancy. Seminars in Oncology 1992 (19, suppl 3): 185 -190

79 Dreicer R, Forest P, Williams RD: A phase II study of 5-fluorouracil (5FU) and alpha interferon in hormone-refractory metastatic prostate cancer. Proc ASCO 1992 (11):205

80 Pigon R, Dillon LP, Hemingway IH, Gearing AJH: Soluble forms of E-selectin, ICAM-1 and VCAM-1 are present in the supernatants of cytokine activated cultured endothelial cells. Biochem Biophys Res Comm 1992 (187):584-589

81 Bulbul MA, Huben RP, Murphy GP: Interferon-ß treatment of metastatic prostate cancer. J Surg Oncol 1986 (33):231-233

82 Kosaki G, Tominaga T, Nakao I, Niitani H: Phase I clinical study of human recombinant interferon gamma on various cancers. Proc 14th Int Cancer Cong, Budapest 1986 (1): Abstr 2485

83 Brunda M, Rosenbaum D, Stern L: Inhibition of experimentally-induced murine metastases by recombinant alpha interferon: correlation between the modulatory effect of interferon treatment. Int J Cancer 1984 (34):421-426

84 Ramani P, Hart IR, Balkwill FR: The effect of interferon on experimental metastases in immunocompetent and immunodeficient mice. Int J Cancer 1986 (37):563-568

85 Horn Y, Zeidman JL, Heller A, Hacohen D, Salzberg S: Serum interferon as a biological marker in malignant tumors. Oncology 1985 (42):164-168

86 Kita M, Nakagawa S, Watanabe H, Kitoh I, Yamaji M, Kishida T, Imanishi J: Interferon production capacity in patients with prostatic cancer: prognostic value. C R Soc Biol 1990 (184):181-186

87 Erlacher L, Gisslinger H, Chott A: Low dosage long-term treatment with recombinant alpha interferon induces in hairy cell leukemia continuous complete remission. Wien Klin Wochenschr 1991 (103):588-590